Practical Guide
to ICP-MS

PRACTICAL SPECTROSCOPY

A SERIES

ADDITIONAL VOLUMES IN PREPARATION

Practical Guide to ICP-MS

Robert Thomas

Scientific Solutions
Gaithersburg, Maryland, U.S.A.

MARCEL DEKKER, INC.   NEW YORK · BASEL

Chemistry Library

Library of Congress Cataloging-in-Publication Data
A catalog record for this book is available from the Library of Congress.

ISBN: 0-8247-5319-4

This book is printed on acid-free paper.

Headquarters
Marcel Dekker, Inc., 270 Madison Avenue, New York, NY 10016, U.S.A.
tel: 212-696-9000; fax: 212-685-4540

Distribution and Customer Service
Marcel Dekker, Inc., Cimarron Road, Monticello, New York 12701, U.S.A.
tel: 800-228-1160; fax: 845-796-1772

Eastern Hemisphere Distribution
Marcel Dekker AG, Hutgasse 4, Postfach 812, CH-4001 Basel, Switzerland
tel: 41-61-260-6300; fax: 41-61-260-6333

World Wide Web
http://www.dekker.com

The publisher offers discounts on this book when ordered in bulk quantities. For more information, write to Special/Professional Marketing at the headquarters address above.

Current printing (last digit):

10 9 8 7 6 5 4 3 2 1

PRINTED IN THE UNITED STATES OF AMERICA

To my ever supportive wife, Donna Marie, and my two precious daughters, Deryn and Glenna.

Foreword

Milestones mark great events: walking on the moon, analyzing rocks on Mars, flying a self-propelled, heavier-than-air machine, using a Bunsen burner for flame atomic spectrometry, and perhaps employing an atmospheric pressure plasma mass spectrometry as an ion source for solution mass spectrometry. Yes, inductively coupled plasma mass spectrometry (ICP-MS) ranks among the milestone inventions of spectrochemical analysis during the 20th century. The great event of ICP-Ms, however, is the enrichment of quantitative ultratrace element and isotope analysis capabilities that has become possible on a daily, routine basis in modern analytical, clinical, forensics, and industrial laboratories. During the past 20 years ICP-MS has grown from R. Sam Houk's Ph.D. research project at the Ames Laboratory on the Iowa State University campus to an invaluable tool fabricated on many continents and applied internationally. Although ICP-MS does not share the universal practicality of the electric light, the laser, or the transistor, it ranks in analytical chemistry along with the development of atomic absorption spectrophotometry, coulometry, dc arc and spark emission spectrography, gravimetry, polarography, and titrimetry.

What can we expect to find in a new technical book, especially one describing ICP-MS in few hundred pages? Do we anticipate a refreshing approach to a well-established topic, answers to unsolved questions, clear insights into complicated problems, astute reviews and critical evaluations of developments, and meaningful consideration of areas for future advancement? We would be satisfied if any of these goals were achieved. Today library bookshelves bear the weight of the writing efforts of numerous recognized researchers and a few practitioners of ICP. Some of these works deserve to stay in the library, while very few others are kept at hand on the analyst's desk, with stained pages and worn bindings as evidence of their heavy use. This volume is intended to be among the latter.

Practical Guide to ICP-MS started as a series of brief tutorial articles ("A Beginner's Guide to ICP-MS") appearing in *Spectroscopy* magazine (Eugene, Oregon; www.spectroscopyonline.com), beginning in April 2001, and it retains the earthy feeling and pragmatism of these monthly contributions. These popular articles were refreshingly straightforward and technically realistic. Presented in an informal style, they reflected the author's years of practical experience on the commercial side of spectroscopic instrumentation and his technical writing skills. Almost immediately I incorporated them into my own spectroscopy teaching programs.

Practical Guide to ICP-MS builds upon this published series. What Robert Thomas has assembled in this volume is 21 chapters that start with basic plasma concepts and ICP-MS instrument component descriptions and conclude with factors to be considered in selecting ICP-MS instruments. Chapters 2 through 16 closely follow the *Spectroscopy* magazines articles I–XII (2001–2002), and Chapter 19 reflects articles XIII and XIV (February 2003). The remaining five chapters comprise others materials, including contamination issues, routine maintenance, prevalent applications areas, comparison with other atomic spectroscopy methods (also adapted from two previously published magazine articles), selection of an ICP-MS system, and contact references.

This is not a handbook describing how to prepare a sample for trace element analysis, perform an ICP-MS measurement. or troubleshoot practical ICP systems. Although these topics urgently need to be addressed, this book is intended to get readers started with ICP-MS. It highlights everything from basic component descriptions and features to guidelines describing where and when using ICP-MS is most appropriately employed. The informal writing style, often in the first person, conveys the author's involvement with ICP product development and his experience with practical applications and makes this text very readable. Consequently, I look forward to seeing this book used in may training programs, classrooms, and analysis laboratories.

Ramon M. Barnes
Director
University Research Institute for Analytical Chemistry
Amherst, Massachusetts, U.S.A.
and
Professor Emeritus
Department of Chemistry
Lederle Graduate Research Center Towers
University of Massachusetts
Amherst, Massachusetts, U.S.A.

Preface

Twenty years after the commercialization of inductively coupled plasma mass spectrometry (ICP-MS) at the Pittsburgh Conference in 1983, approximately 5,000 systems have been installed worldwide. If this is compared with another rapid multielement technique, inductively coupled plasma optical emission spectrometry (ICP-OES), first commercialized in 1974, the difference is quite significant. As of 1994, 20 years after ICP-OES was introduced, about 12,000 units had been sold, and if this is compared with the same time period for which ICP-MS has been available the difference is even more staggering. From 1983 to the present day, approximately 25,000 ICP-OES systems have been installed—about 5 times more than the number of ICP-MS systems. If the comparison is made with all atomic spectroscopy instrumentation (ICP-MS, ICP-OES, Electrothermal Atomization [ETA], and flame atomic absorption [FAA]), the annual sales for ICP-MS are less than 7% of the total AS market—500 units compared with approximately 7000 AS systems. It's even more surprising when one considers that ICP-MS offers so much more than the other techniques, including superb detection limits, rapid multielement analysis and isotopic measurement capabilities.

ICP-MS: RESEARCH OR ROUTINE?

Clearly, one of the many reasons that ICP-MS has not become more popular is its relatively high price-tag—an ICP mass spectrometer typically cost 2 times more than ICP-OES and 3 times more than ETA. But in a competitive world, the street price of an ICP-MS system is much closer to a top-of-the-line ICP-OES with sampling accessories or an ETA system that has all the bells and whistles on it. So if ICP-MS is not significantly more expensive than ICP-OES and ETA, why hasn't it been more widely accepted by the analytical community? The answer may lie in the fact that it is still considered a complicated research-type technique, requiring a very skilled person to operate it.

Manufacturers of ICP-MS equipment are constantly striving to make the systems easier to operate, the software easier to use and the hardware easier to maintain, but even after 20 years, it is still not perceived as a mature, routine tool like flame AA or ICP-OES. This might be partially true because of the relative complexity of the instrumentation. However, could the dominant reason for this misconception be the lack of availability of good literature explaining the basic principles and application benefits of ICP-MS, in a way that is compelling and easy to understand for a novice who has limited knowledge of the technique? There are some excellent textbooks (1–3) and numerous journal papers (4,5,6) available describing the fundamentals, but they are mainly written or edited by academics who are not approaching the subject from a practical perspective. For this reason, they tend to be far too heavily biased toward basic principles and less toward how ICP-MS is being applied in the real-world.

PRACTICAL BENEFITS

There is no question that the technique needs to be presented in a more practical way, in order to make routine analytical laboratories more comfortable with it. Unfortunately, the publisher of the *Dummies* series has not yet found a mass market for a book on ICP-MS. This is being a little facetious, of course, but, from the limited number of ICP-MS reference books available today, it is clear that a practical guide is sadly lacking. This was most definitely the main incentive for writing the book. However, it was also felt that to paint a complete picture for someone who is looking to invest to ICP-MS, it was very important to compare its capabilities with those of other common trace element techniques, such as FAA, ETA, and ICP-OES, focusing on such criteria as elemental range, detection capability, sample throughput, analytical working range, interferences, sample preparation, maintenance issues, operator skill level, and running costs. This will enable the reader to relate the benefits of ICP-MS to those of other more familiar atomic spectroscopy instrumentation. In addition, in order to fully understand its practical capabilities, it is important to give an overview of the most common applications currently being carried out by ICP-MS and its sampling accessories, to give a flavor of the different industries and markets that are benefiting from the technique's enormous potential. And finally, for those who might be interested in purchasing the technique, the book concludes with a chapter on the most important selection criteria. This is critical ingredient in presenting ICP-MS to a novice, because there is very little information in the public domain to help someone carry out an evaluation of commercial instrumentation. Very often, people go into this evaluation process completely unprepared and as a result may end up with an instrument that is not ideally suited for their needs.

The main objective is to make ICP-MS a little more compelling to purchase and ultimately open up its potential to the vast majority of the trace element community who have not yet realized the full benefits of its capabilities. With this in mind, please feel free to come in and share one person's view of ICP-MS and its applications.

ACKNOWLEDGMENTS

I have been working in the field of ICP mass spectrometry for almost 20 years and realized that, even though numerous publications were available, no textbooks were being written specifically for beginners with a very limited knowledge of the technique. I came to the conclusion that the only way this was going to happen was to write it myself. I set myself the objective of putting together a reference book that could be used by both analytical chemists and senior management who were experienced in the field of trace metals analysis, but only had a basic understanding of ICP-MS and the benefits it had to offer. This book represents the conclusion of that objective. So now after two years of hard work, I would like to take this opportunity to thank some of the people and organizations that have helped me put the book together. First, I would like to thank the editorial staff of Spectroscopy magazine, who gave me the opportunity to write a monthly tutorial on ICP-MS back in the spring of 2001, and also allowed me to use many of the figures from the series-this was most definitely the spark I needed to start the project. Second, I would like to thank all the manufacturers of ICP-MS instrumentation, equipment, accessories, consumables, calibration standards and reagents, who supplied me with the information, data, drawings and schematics etc. It would not have been possible without their help. Third, I would like to thank Dr. Ramon Barnes, Director of the University Research Institute for Analytical Chemistry and organizer/chairman of the Winter Conference on Plasma Spectrochemistry for the kind and complimentary words he wrote in the Foreword— they were very much appreciated. Finally, I would like to thank my truly inspirational wife, Donna Marie, for allowing me to take up full-time writing four years ago and particularly for her encouragement over the past two years while writing the book. Her support was invaluable. And I mustn't forget my two precious daughters, Glenna and Deryn, who kept me entertained and amused, especially during the final proofing/indexing stage when I thought I would never get the book finished. I can still hear their words of wisdom, "Dad, it's only a book."

FURTHER READING

1. Inductive by Coupled Plasma Mass Spectrometry: A. Montasser, George Washington University, *Wiley-VCH*, New York, 1998.

2. Handbook of Inductively Coupled Plasma Mass Spectrometry: K. E. Jarvis, A. L. Gray and R. S. Houk, *Blackie*, Glasgow, 1992.

3. Inorganic Mass Spectrometry, F. Adams, R. Gijbels, R. Van Grieken, University of Antwerp, *Wiley and Sons*, New York, 1988.

4. R.S. Houk, V. A. Fassel and H. J. Svec, *Dynamic Mass Spec.* 6, 234, 1981.

5. A.R. Date and A.L. Gray, *Analyst*, 106, 1255, 1981.

6. D. J. Douglas and J. B. French, *Analytical Chemistry*, 53, 37, 1982.

Robert Thomas

Contents

Practical Guide to ICP-MS

1

An Overview of ICP–Mass Spectrometry

Inductively coupled plasma mass spectrometry (ICP-MS) not only offers extremely low detection limits in the sub parts per trillion (ppt) range, but also enables quantitation at the high parts per million (ppm) level. This unique capability makes the technique very attractive compared to other trace metal techniques such as electrothermal atomization (ETA), which is limited to determinations at the trace level, or flame atomic absorption (FAA) and inductively coupled plasma optical emission spectroscopy (ICP-OES), which are traditionally used for the detection of higher concentrations. In Chapter 1, we will present an overview of ICP-MS and explain how its characteristic low detection capability is achieved.

Inductively coupled plasma mass spectrometry (ICP-MS) is undoubtedly the fastest-growing trace element technique available today. Since its commercialization in 1983, approximately 5000 systems have been installed worldwide, carrying out many varied and diverse applications. The most common ones, which represent approximately 80% of the ICP-MS analyses being carried out today, include environmental, geological, semiconductor, biomedical, and nuclear application fields. There is no question that the major reason for its unparalleled growth is its ability to carry out rapid multielement determinations at the ultra trace level. Even though it can broadly determine the same suite of elements as other atomic spectroscopical techniques, such as flame atomic absorption (FAA), electrothermal atomization (ETA), and inductively coupled plasma optical emission spectroscopy (ICP-OES), ICP-MS has clear advantages in its multielement characteristics, speed of analysis, detection limits, and isotopic capability. Figure 1.1 shows approximate detection limits of all the elements that can be detected by ICP-MS, together with their isotopic abundance.

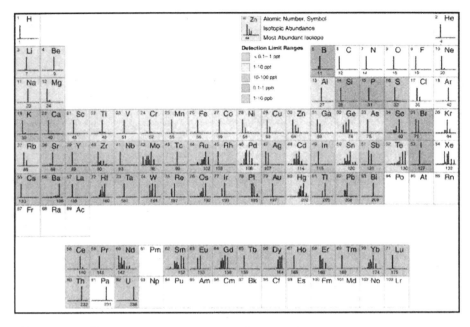

FIGURE 1.1 Detection limit capability of ICP-MS. (Courtesy of Perkin-Elmer Life and Analytical Sciences.)

PRINCIPLES OF OPERATION

There are a number of different ICP-MS designs available today, which share many similar components, such as nebulizer, spray chamber, plasma torch, and detector, but can differ quite significantly in the design of the interface, ion focusing system, mass separation device, and vacuum chamber. Instrument hardware will be described in greater detail in the subsequent chapters, but first let us start by giving an overview of the principles of operation of ICP-MS. Figure 1.2 shows the basic components that make up an ICP-MS system. The sample, which usually must be in a liquid form, is pumped at 1 mL/min, usually with a peristaltic pump into a nebulizer, where it is converted into a fine aerosol with argon gas at about 1 L/min. The fine droplets of the aerosol, which represent only 1–2% of the sample, are separated from larger droplets by means of a spray chamber. The fine aerosol then emerges from the exit tube of the spray chamber and is transported into the plasma torch via a sample injector.

It is important to differentiate the roll of the plasma torch in ICP-MS compared to ICP-OES. The plasma is formed in exactly the same way, by

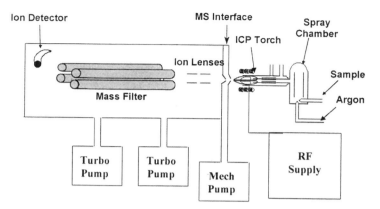

FIGURE 1.2   Basic instrumental components of an ICP mass spectrometer.

the interaction of an intense magnetical field [produced by radiofrequency (RF) passing through a copper coil] on a tangential flow of gas (normally argon), at about 15 L/min flowing through a concentrical quartz tube (torch). This has the effect of ionizing the gas and, when seeded with a source of electrons from a high-voltage spark, forms a very-high-temperature plasma discharge (~10,000 K) at the open end of the tube. However, this is where the similarity ends. In ICP-OES, the plasma, which is normally vertical, is used to generate photons of light, by the excitation of electrons of a ground-state atom to a higher energy level. When the electrons "fall" back to ground state, wavelength-specific photons are emitted, which are characteristic of the element of interest. In ICP-MS, the plasma torch, which is positioned horizontally, is used to generate positively charged ions and not photons. In fact, every attempt is made to stop the photons from reaching the detector because they have the potential to increase signal noise. It is the production and the detection of large quantities of these ions that give ICP-MS its characteristic low parts per trillion (ppt) detection capability—about three to four orders of magnitude better than ICP-OES.

Once the ions are produced in the plasma, they are directed into the mass spectrometer via the interface region, which is maintained at a vacuum of 1–2 Torr with a mechanical roughing pump. This interface region consists of two metallic cones (usually nickel), called the sampler and a skimmer cone, each with a small orifice (0.6–1.2 mm) to allow the ions to pass through to the ion optics, where they are guided into the mass separation device.

The interface region is one of the most critical areas of an ICP mass spectrometer because the ions must be transported efficiently and with electrical integrity from the plasma, which is at atmospheric pressure (760 Torr) to the mass spectrometer analyzer region at approximately 10^{-6} Torr. Unfortunately, there is capacitive coupling between the RF coil and the plasma, producing a potential difference of a few hundred volts. If this were not eliminated, it would have resulted in an electrical discharge (called a secondary discharge or pinch effect) between the plasma and the sampler cone. This discharge increases the formation of interfering species and also dramatically affects the kinetic energy of the ions entering the mass spectrometer, making optimization of the ion optics very erratic and unpredictable. For this reason, it is absolutely critical that the secondary charge is eliminated by grounding the RF coil. There have been a number of different approaches used over the years to achieve this, including a grounding strap between the coil and the interface, balancing the oscillator inside the RF generator circuitry, a grounded shield or plate between the coil and the plasma torch, or the use of a double interlaced coil where RF fields go in opposing directions. They all work differently, but basically achieve a similar result, which is to reduce or to eliminate the secondary discharge.

Once the ions have been successfully extracted from the interface region, they are directed into the main vacuum chamber by a series of electrostatic lens, called ion optics. The operating vacuum in this region is maintained at about 10^{-3} Torr with a turbomolecular pump. There are many different designs of the ion optical region, but they serve the same function, which is to electrostatically focus the ion beam toward the mass separation device, while stopping photons, particulates, and neutral species from reaching the detector.

The ion beam containing all the analytes and matrix ions exits the ion optics and now passes into the heart of the mass spectrometer—the mass separation device, which is kept at an operating vacuum of approximately 10^{-6} Torr with a second turbomolecular pump. There are many different mass separation devices, all with their strengths and weaknesses. Four of the most common types are discussed in this book—quadrupole, magnetic sector, time of flight, and collision/reaction cell technology—but they basically serve the same purpose, which is to allow analyte ions of a particular mass-to-charge ratio through to the detector and to filter out all the nonanalyte, interfering, and matrix ions. Depending on the design of the mass spectrometer, this is either a scanning process, where the ions arrive at the detector in a sequentially manner, or a simultaneous process, where the ions are either sampled or detected at the same time.

The final process is to convert the ions into an electrical signal with an ion detector. The most common design used today is called a discrete dynode detector, which contain a series of metal dynodes along the length of the detector. In this design, when the ions emerge from the mass filter, they impinge on the first dynode and are converted into electrons. As the electrons are attracted to the next dynode, electron multiplication takes place, which results in a very high steam of electrons emerging from the final dynode. This electronic signal is then processed by the data handling system in the conventional way and then converted into analyte concentration using ICP-MS calibration standards. Most detection systems used can handle up to eight orders of dynamic range, which means that they can be used to analyze samples from ppt levels, up to a few hundred parts per million (ppm).

It is important to emphasize that because of the enormous interest in the technique, most ICP-MS instrument companies have very active R&D programs in place, in order to get an edge in a very competitive marketplace. This is obviously very good for the consumer because not only does it drive down instrument prices, but also the performance, applicability, usability, and flexibility of the technique are improved at an alarming rate. Although this is extremely beneficial for the ICP-MS user community, it can pose a problem for a textbook writer who is attempting to present a snapshot of instrument hardware and software components at a particular moment in time. Hopefully, I have struck the right balance in not only presenting the fundamental principles of ICP-MS to a beginner, but also making them aware of what the technique is capable of achieving and where new developments might be taking it.

2

Principles of Ion Formation

Chapter 2 gives a brief overview of the fundamental principle used in inductively coupled plasma mass spectrometry (ICP-MS)—the use of a high-temperature argon plasma to generate positive ions. The highly energized argon ions that make up the plasma discharge are used to first produce analyte ground state atoms from the dried sample aerosol, and then to interact with the atoms to remove an electron and to generate positively charged ions, which are then steered into the mass spectrometer for detection and measurement.

In inductively coupled plasma mass spectrometry the sample, which is usually in liquid form, is pumped into the sample introduction system, comprising a spray chamber and a nebulizer. It emerges as an aerosol, where it eventually finds its way via a sample injector into the base of the plasma. As it travels through the different heating zones of the plasma torch, it is dried, vaporized, atomized, and ionized. During this time, the sample is transformed from a liquid aerosol to solid particles, then into gas. When it finally arrives at the analytical zone of the plasma, at approximately 6000–7000 K, it exists as ground state atoms and ions, representing the elemental composition of the sample. The excitation of the outer electron of a ground state atom to produce wavelength-specific photons of light is the fundamental basis of atomic emission. However, there is also enough energy in the plasma to remove an electron from its orbital to generate a free ion. The energy available in an argon plasma is ~15.8 eV, which is high enough to ionize most of the elements in the periodic table (the majority have first ionization potentials in the order of 4–12 eV). It is the generation, transportation, and detection of significant numbers of positively charged ions that give ICP-MS its characteristic ultra trace detection capabilities. It is also important to mention that although ICP-MS is predominantly used for the detection of positive ions, negative ions (e.g., halogens) are also produced in the plasma. However, because the extraction and the transportation of negative ions are different from that of

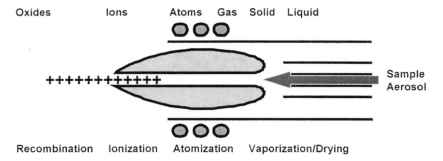

Oxides   Ions   Atoms Gas Solid Liquid

Sample Aerosol

Recombination Ionization Atomization Vaporization/Drying

FIGURE 2.1 Generation of positively charged ions in the plasma.

positive ions, most commercial instruments are not designed to measure them. The process of the generation of positively charged ions in the plasma is conceptually shown in greater detail in Figure 2.1.

ION FORMATION

The actual process of conversion of a neutral ground state atom to a positively charged ion is shown in Figures 2.2 and 2.3. Figure 2.2 shows a

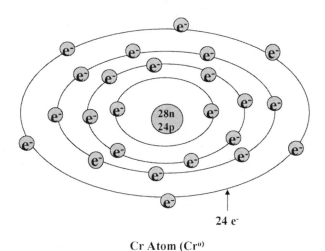

24 e⁻

Cr Atom (Cr⁰)

FIGURE 2.2 Simplified schematic of a chromium ground sate atom (Cr⁰).

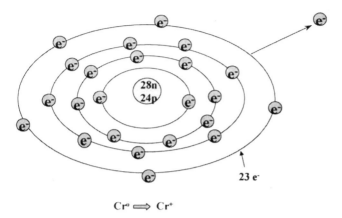

$$Cr^0 \Longrightarrow Cr^+$$

FIGURE 2.3 Conversion of a chromium ground state atom (Cr^0) to an ion (Cr^+).

very simplistic view of the chromium atom Cr^0, consisting of a nucleus with 24 protons (p^+) and 28 neutrons (n), surrounded by 24 orbiting electrons (e^-). (It must be emphasized that this is not meant to be an accurate representation of the electrons' shells and subshells, but just a conceptual explanation for the purpose of clarity.) From this, we can say that the atomic number of chromium is 24 (number of protons) and its atomic mass is 52 (number of protons + neutrons).

If energy is then applied to the chromium ground sate atom in the form of heat from a plasma discharge, one of the orbiting electrons will be stripped off the outer shell. This will result in only 23 electrons left orbiting the nucleus. Because the atom has lost a negative charge (e^-), but still has 24 protons (p^+) in the nucleus, it is converted into an ion with a net positive charge. It still has an atomic mass of 52 and an atomic number of 24, but is now a positively charged ion and not a neutral ground state atom. This process is shown in Figure 2.3.

NATURAL ISOTOPES

This is a very basic look at the process because most elements occur in more than one form (isotope). In fact, chromium has four naturally occurring isotopes, which means that the chromium atom exists in four different forms, all with the same atomic number of 24 (number of protons) but with different atomic masses (number of neutrons).

To make this a little easier to understand, let us take a closer look at an element such as copper, which only has two different isotopes—one with an

TABLE 2.1 Breakdown of the Atomic Structure of
Copper Isotopes

	^{63}Cu	^{65}Cu
Number of protons (p^+)	29	29
Number electrons (e^-)	29	29
Number of neutrons (n)	34	36
Atomic mass ($p^+ + n$)	63	65
Atomic number (p^+)	29	29
Natural abundance (%)	69.17	30.83
Nominal atomic weight	63.55^a	

[a] The nominal atomic weight of copper is calculated using the formula: $0.6917n\,(^{63}Cu) + 0.3083n\,(^{65}Cu) + p^+$, and is referenced to the atomic weight of carbon.

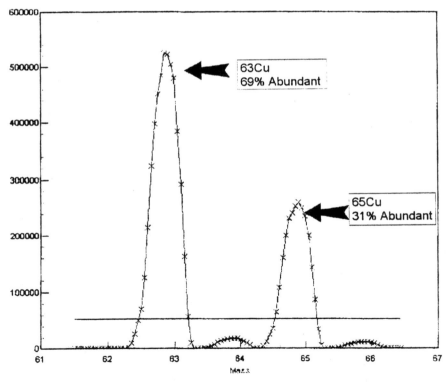

FIGURE 2.4 Mass spectra of the two copper isotopes—$^{63}Cu^+$ and $^{65}Cu^+$.

Isotope		%		%		%
1	H	99.985				
2	H	0.015				
3			He	0.000137		
4			He	99.999863		
5						
6					Li	7.5
7					Li	92.5
8						
9	Be	100				
10			B	19.9		
11			B	80.1		
12					C	98.90
13					C	1.10
14	N	99.643				
15	N	0.366				
16			O	99.762		
17			O	0.038		
18			O	0.200		
19					F	100
20	Ne	90.48				
21	Ne	0.27				
22	Ne	9.25				
23			Na	100		
24					Mg	78.99
25					Mg	10.00
26					Mg	11.01
27	Al	100				
28			Si	92.23		
29			Si	4.67		
30			Si	3.10		
31					P	100
32	S	95.02				
33	S	0.75				
34	S	4.21				
35			Cl	75.77		
36	S	0.02			Ar	0.337
37			Cl	24.23		
38					Ar	0.063
39	K	93.2581				
40	K	0.0117	Ca	96.941	Ar	99.600
41	K	6.7302				
42			Ca	0.647		
43			Ca	0.135		
44			Ca	2.086		
45					Sc	100
46	Ti	8.0	Ca	0.004		
47	Ti	7.3				
48	Ti	73.8	Ca	0.187		
49	Ti	5.5				
50	Ti	5.4	V	0.250	Cr	4.345
51			V	99.750		
52					Cr	83.789
53					Cr	9.501
54	Fe	5.8			Cr	2.365
55			Mn	100		
56	Fe	91.72				
57	Fe	2.2				
58	Fe	0.28			Ni	68.077
59			Co	100		
60					Ni	26.223

Isotope		%		%		%
61					Ni	1.140
62					Ni	3.634
63	Cu	69.17				
64			Zn	48.6	Ni	0.926
65	Cu	30.83				
66			Zn	27.9		
67			Zn	4.1		
68			Zn	18.8		
69					Ga	60.108
70	Ge	21.23	Zn	0.6		
71					Ga	39.892
72	Ge	27.66				
73	Ge	7.73				
74	Ge	35.94	Se	0.89		
75					As	100
76	Ge	7.44	Se	9.36		
77			Se	7.63		
78	Kr	0.35	Se	23.78		
79					Br	50.69
80	Kr	2.25	Se	49.61		
81					Br	49.31
82	Kr	11.6	Se	8.73		
83	Kr	11.5				
84	Kr	57.0	Sr	0.56		
85					Rb	72.165
86	Kr	17.3	Sr	9.86		
87			Sr	7.00	Rb	27.835
88			Sr	82.58		
89					Y	100
90	Zr	51.45				
91	Zr	11.22				
92	Zr	17.15	Mo	14.84		
93					Nb	100
94	Zr	17.38	Mo	9.25		
95			Mo	15.92		
96	Zr	2.80	Mo	16.68	Ru	5.52
97			Mo	9.55		
98			Mo	24.13	Ru	1.88
99					Ru	12.7
100			Mo	9.63	Ru	12.6
101					Ru	17.0
102	Pd	1.02			Ru	31.6
103			Rh	100		
104	Pd	11.14			Ru	18.7
105	Pd	22.33				
106	Pd	27.33	Cd	1.25		
107					Ag	51.839
108	Pd	26.46	Cd	0.89		
109					Ag	48.161
110	Pd	11.72	Cd	12.49		
111			Cd	12.80		
112	Sn	0.97	Cd	24.13		
113			Cd	12.22	In	4.3
114	Sn	0.65	Cd	28.73		
115	Sn	0.34			In	95.7
116	Sn	14.53	Cd	7.49		
117	Sn	7.68				
118	Sn	24.23				
119	Sn	8.59				
120	Sn	32.59	Te	0.096		

Isotope		%		%		%
121					Sb	57.36
122	Sn	4.63	Te	2.603		
123			Te	0.908	Sb	42.64
124	Sn	5.79	Te	4.816	Xe	0.10
125			Te	7.139		
126			Te	18.95	Xe	0.09
127	I	100				
128			Te	31.69	Xe	1.91
129					Xe	26.4
130	Ba	0.106	Te	33.80	Xe	4.1
131					Xe	21.2
132	Ba	0.101			Xe	26.9
133			Cs	100		
134	Ba	2.417			Xe	10.4
135	Ba	6.592				
136	Ba	7.854	Ce	0.19	Xe	8.9
137	Ba	11.23				
138	Ba	71.70	Ce	0.25	La	0.0902
139					La	99.9098
140			Ce	88.48		
141					Pr	100
142	Nd	27.13	Ce	11.08		
143	Nd	12.18				
144	Nd	23.80	Sm	3.1		
145	Nd	8.30				
146	Nd	17.19				
147			Sm	15.0		
148	Nd	5.76	Sm	11.3		
149			Sm	13.8		
150	Nd	5.64	Sm	7.4		
151					Eu	47.8
152	Gd	0.20	Sm	26.7		
153					Eu	52.2
154	Gd	2.18	Sm	22.7		
155	Gd	14.80				
156	Gd	20.47	Dy	0.06		
157	Gd	15.65				
158	Gd	24.84	Dy	0.10		
159					Tb	100
160	Gd	21.86	Dy	2.34		
161			Dy	18.9		
162	Er	0.14	Dy	25.5		
163			Dy	24.9		
164	Er	1.61	Dy	28.2		
165					Ho	100
166	Er	33.6				
167	Er	22.95				
168	Er	26.8	Yb	0.13		
169					Tm	100
170	Er	14.9	Yb	3.05		
171			Yb	14.3		
172			Yb	21.9		
173			Yb	16.12		
174			Yb	31.8	Hf	0.162
175	Lu	97.41				
176	Lu	2.59	Yb	12.7	Hf	5.206
177					Hf	18.606
178					Hf	27.297
179					Hf	13.629
180	Ta	0.012	W	0.13	Hf	35.100

Isotope		%		%		%
181	Ta	99.988				
182			W	26.3		
183			W	14.3		
184	Os	0.02	W	30.67		
185					Re	37.40
186	Os	1.58	W	28.6		
187	Os	1.6			Re	62.60
188	Os	13.3				
189	Os	16.1				
190	Os	26.4			Pt	0.01
191			Ir	37.3		
192	Os	41.0			Pt	0.79
193			Ir	62.7		
194					Pt	32.9
195					Pt	33.8
196	Hg	0.15			Pt	25.3
197			Au	100		
198	Hg	9.97			Pt	7.2
199	Hg	16.87				
200	Hg	23.10				
201	Hg	13.18				
202	Hg	29.86				
203					Tl	29.524
204	Hg	6.87	Pb	1.4		
205					Tl	70.476
206			Pb	24.1		
207			Pb	22.1		
208			Pb	52.4		
209	Bi	100				
210						
211						
212						
213						
214						
215						
216						
217						
218						
219						
220						
221						
222						
223						
224						
225						
226						
227						
228						
229						
230						
231	Pa	100				
232	Th	100				
233						
234	U	0.0055				
235	U	0.7200				
236						
237						
238	U	99.2745				

"Isotopic Compositions of the Elements 1989" Pure Appl. Chem., Vol. 63, No. 7, pp. 991-1002, 1991 © 1991 IUPAC

FIGURE 2.5 Relative abundance of the naturally occurring isotopes of elements. (From Ref. 1.)

atomic mass of 63 (^{63}Cu) and another with an atomic mass of 65 (^{65}Cu). They both have the same number of protons and electrons, but differ in the number of neutrons in the nucleus. The natural abundances of ^{63}Cu and ^{65}Cu are 69.1% and 30.9%, respectively, which gives copper a nominal atomic mass of 63.55—the value you see for copper in atomic weight reference tables. Details of the atomic structure of the two copper isotopes are shown in Table 2.1.

When a sample containing naturally occurring copper is introduced into the plasma, two different ions of copper, $^{63}Cu^+$ and $^{65}Cu^+$, are produced, which generate two different mass spectra—one at mass 63 and another at mass 65. This can be seen in Figure 2.4, which is an actual ICP-MS spectral scan of a sample containing copper, showing a peak for the $^{63}Cu^+$ ion on the left, which is 69.17% abundant, and a peak for $^{65}Cu^+$ at 30.83% abundance, on the right. You can also see small peaks for two Zn isotopes at mass 64 ($^{64}Zn^+$) and mass 66 ($^{66}Zn^+$). (Zn has a total of five isotopes at masses 64, 66, 67, 68, and 70.) In fact, most elements have at least two or three isotopes, and many elements, including zinc and lead, have four or more isotopes. Figure 2.5 is a chart showing the relative abundance of the naturally occurring isotopes of all elements.

FURTHER READING

1. Isotopic composition of the elements. Pure Appl Chem 1991; 63(7):991–1002. (UIPAC).

3

Sample Introduction

Chapter 3 examines one of the most critical areas of the instrument—the sample introduction system. It will discuss the fundamental principles of converting a liquid into a fine-droplet aerosol suitable for ionization in the plasma, together with an overview of the different types of commercially available nebulizers and spray chambers.

The majority of ICP-MS applications carried out today involve the analysis of liquid samples. Even though the technique has been adapted over the years to handle solids and slurries, it was developed in the early 1980s primarily to analyze solutions. There are many different ways of introducing a liquid into an ICP mass spectrometer, but they all basically achieve the same result, and that is to generate a fine aerosol of the sample, so it can be efficiently ionized in the plasma discharge. The sample introduction area has been called the "Achilles Heel" of ICP-MS, because it is considered the weakest component of the instrument—with only 1–2% of the sample finding its way into the plasma [1]. Although there has recently been much improvement in this area, the fundamental design of an ICP-MS sample introduction system has not dramatically changed since the technique was first introduced in 1983.

Before we discuss the mechanics of aerosol generation in greater detail, let us look at the basic components of a sample introduction system. Figure 3.1 shows the proximity of the sample introduction area relative to the rest of the ICP mass spectrometer, while Figure 3.2 represents a more detailed view showing the individual components.

The mechanism of introducing a liquid sample into an analytical plasma can be considered as two separate events—aerosol generation using a nebulizer and droplet selection by way of a spray chamber [2].

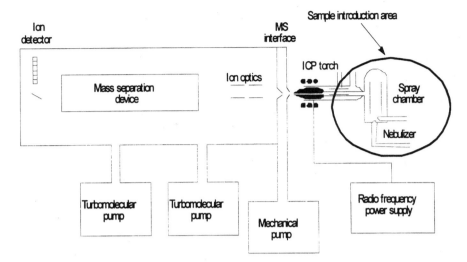

FIGURE 3.1   Location of the ICP-MS sample introduction area.

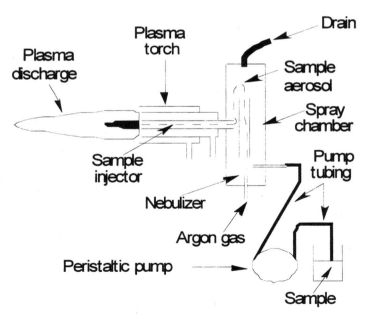

FIGURE 3.2   More detailed view of the ICP-MS sample introduction area.

AEROSOL GENERATION

As previously mentioned, the main function of the sample introduction system is to generate a fine aerosol of the sample. It achieves this with a nebulizer and a spray chamber. The sample is normally pumped at about 1 mL/min via a peristaltic pump into the nebulizer. A peristaltic pump is a small pump with lots of mini-rollers that all rotate at the same speed. The constant motion and pressure of the rollers on the pump tubing feeds the sample through to the nebulizer. The benefit of a peristaltic pump is that it ensures a constant flow of liquid, irrespective of differences in viscosity between samples, standards, and blanks. Once the sample enters the nebulizer, the liquid is then broken up into a fine aerosol by the pneumatic action of a flow of gas (~1 L/min) "smashing" the liquid into tiny droplets, very similar to the spray mechanism of a can of deodorant. It should be noted that although pumping the sample is the most common approach to introduce the sample, some pneumatic nebulizers such as the concentric design do not necessitate the use of a pump, because they rely on the natural "venturi effect" of the positive pressure of the nebulizer gas to suck the sample through the tubing. Solution nebulization is conceptually represented in Figure 3.3, which shows aerosol generation using a crossflow-designed nebulizer.

DROPLET SELECTION

Because the plasma discharge is not very efficient at dissociating large droplets, the function of the spray chamber is primarily to allow only the small droplets to enter the plasma. Its secondary purpose is to smooth out pulses that occur during nebulization process, due mainly to the peristaltic pump. There are a number of different ways of ensuring that only the small droplets get through, but the most common way is to use a double-pass spray

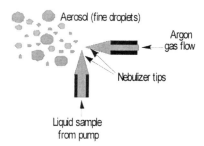

FIGURE 3.3 Conceptual representation of aerosol generation using a nebulizer.

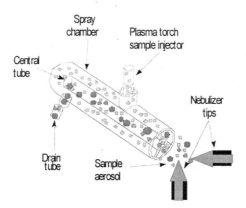

FIGURE 3.4   Simplified representation of the separation of large droplets from the fine droplets in the spray chamber.

chamber, where the aerosol emerges from the nebulizer and is directed into a central tube running the whole length of the chamber. The droplets travel the length of this tube, where the large droplets (greater than ~10 μm in diameter) will fall out by gravity and exit through the drain tube at the end of the spray chamber. The fine droplets (<10 μm diameter) then pass between the outer wall and the central tube where they eventually emerge from the spray chamber and transported into the sample injector of the plasma torch [3]. Although there are many different designs available, the spray chamber's main function is to allow only the smallest droplets into the plasma for dissociation, atomization, and finally ionization of the sample's elemental components. A simplified schematic of this process is represented in Figure 3.4.

Let us now look at the different nebulizer and spray chamber designs that are most commonly used in ICP-MS. We cannot cover every available type, because over the past few years, a huge market has developed for application-specific, customized sample introduction components. This has, in fact, generated an industry of small OEM (Other Equipment Manufacturers) companies that manufacture parts for instrument companies as well as sell directly to ICP-MS users.

NEBULIZERS

By far, the most common design used for ICP-MS is the pneumatic nebulizer, which uses mechanical forces of a gas flow (normally argon at a pressure of 20–30 psi) to generate the sample aerosol. Some of the most popular designs of pneumatic nebulizer include the concentric, microconcentric, microflow, and crossflow. They are usually made from glass, but other nebulizer materials, such as various kinds of polymers, are becoming more popular, particularly for highly corrosive samples and specialized applications. It should be emphasized at this point that nebulizers designed for use with ICP-OES are far from ideal for use with ICP-MS. This is the result of a limitation in total dissolved solids (TDS) that can be put into the ICP-MS interface area. Because the orifice size of the sampler and skimmer cones used in ICP-MS are so small (~0.6–1.2 mm), the matrix components must be generally kept below 0.2%, although higher concentrations of some matrices can be tolerated (refer to Chapter 5 on the "Interface Region") [4]. This means that general-purpose ICP-OES nebulizers that are designed to aspirate 1–2% dissolved solids, or high solids nebulizers such as the Babbington, V-groove, or cone-spray, which are designed to handle up to 20% dissolved solids, are not ideally suited to analyze solutions by ICP-MS. However, if slurries are being attempted by ICP-MS, as long as the particle sizes is kept below <10 μm in diameter, these types of nebulizers can be very useful [5]. The most common of the pneumatic nebulizers used in commercial ICP mass spectrometers are the concentric and crossflow design. The concentric design is more suitable for clean samples, while the crossflow is generally more tolerant to samples containing higher solids and/or particulate matter.

Concentric Design

In the concentric nebulizer, the solution is introduced through a capillary tube to a low-pressure region created by a gas flowing rapidly past the end of the capillary. The low pressure and high-speed gas combine to break up the solution into an aerosol, which forms at the open end of the nebulizer tip. This is shown in greater detail in Figure 3.5.

Concentric pneumatic nebulizers can give excellent sensitivity and stability, particularly with clean solutions. However, the small orifices can be plagued by blockage problems, especially if large numbers of heavy-matrix samples are being aspirated.

Crossflow Design

For samples that contain a heavier matrix or maybe small amounts of undissolved matter, the crossflow design is probably the best option. With this design, the argon gas is directed at right angles to the tip of a capillary tube, in contrast to the concentric, where the gas flow is parallel to the capillary. The solution is either drawn up through the capillary tube via the pressure created by the high-speed gas flow, or as is most common with crossflow

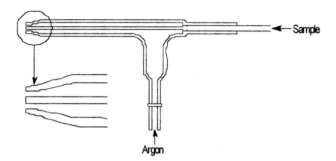

FIGURE 3.5 Typical concentric nebulizer. (Courtesy of Meinhard Glass Products.)

nebulizers, forced through the tube with a peristaltic pump. In either case, contact between the high-speed gas and the liquid stream causes the liquid to break up into an aerosol. Crossflow nebulizers are generally not as efficient as concentric nebulizers at creating the very small droplets needed for ICP-MS analyses. However, the larger-diameter liquid capillary and longer distance between liquid and gas injectors reduces clogging problems. Many analysts feel that the small penalty paid in analytical sensitivity and precision, compared to concentric nebulizers, is compensated by the fact that they are far more rugged for routine use. A cross section of a crossflow nebulizer is shown in Figure 3.6.

Microflow Design

A new breed of nebulizers is being developed for ICP-MS called microflow or high-efficiency nebulizers, which are designed to operate at much lower sample flows. While conventional nebulizers have a sample uptake rate of

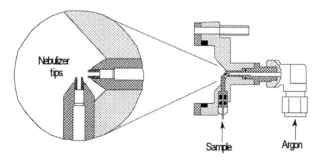

FIGURE 3.6 Schematic of a crossflow nebulizer. (Courtesy of Perkin-Elmer Life and Analytical Sciences.)

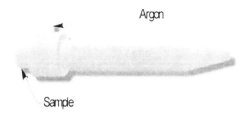

Argon

Sample

FIGURE 3.7   A PFA microflow concentric nebulizer. (Courtesy of Elemental Scientific Inc.)

about 1 mL/min, microflow and high-efficiency nebulizers typically run at less than 0.1 mL/min. They are based on the concentric principal, but usually operate at higher gas pressure to accommodate the lower sample flow rates. The extremely low uptake rate makes them ideal for applications where sample volume is limited or where the sample/analyte is prone to sample introduction memory effects. The additional benefit of this design is that it produces an aerosol with smaller droplets and, as a result, is generally more efficient than a conventional concentric nebulizer.

These nebulizers and their components are typically constructed from polymer materials, such as polytetrafluoroethylene (PTFE), perfluoroalkoxy (PFA), or polyvinylfluoride (PVF), although some designs are available in quartz. The excellent corrosion resistance of the ones made from polymers means they have naturally low blank levels. This characteristic, together with their ability to handle small sample volumes found in applications such as vapor phase decomposition (VPD), makes them an ideal choice for semiconductor laboratories that are carrying out ultratrace element analysis [6]. A microflow nebulizer made from PFA is shown in Figure 3.7.

The disadvantage of a microconcentric nebulizer is that it is not very tolerant to high concentrations of dissolved solids or suspended particles. Their high efficiency means that most of the sample make it into the plasma and, as a result, can cause more severe matrix suppression problems. In addition, the higher dissolved solids going through the interface has the potential to cause cone blockage problems over extended periods of operation. For these reasons, they have been found to be most applicable for the analysis of samples containing low levels of dissolved solids.

SPRAY CHAMBERS

Let us now turn our attention to spray chambers. There are basically three designs that are used in commercial ICP-MS instrumentation—Double Pass, Cyclonic, and Impact Bead spray chambers. The double pass is by far the most common, with the cyclonic type rapidly gaining in popularity. The impact bead design, which was first developed for flame AA, is also an option on some ICP-MS systems. As mentioned earlier, the function of the spray chamber is to reject the larger aerosol droplets and also to smooth out nebulization pulses produced by the peristaltic pump. In addition, some ICP-MS spray chambers are externally cooled (typically to 2–5°C) for thermal stability of the sample and to minimize the amount of solvent going into the plasma. This can have a number of beneficial effects, depending on the application, but the main benefits are reduction of oxide species and the ability to aspirate organic solvents.

Double Pass

By far, the most common design of double-pass spray chamber is the Scott design, which selects the small droplets by directing the aerosol into a central tube. The larger droplets emerge from the tube and by gravity, exit the spray chamber via a drain tube. The liquid in the drain tube is kept at positive pressure (usually by way of a loop), which forces the small droplets back between the outer wall and the central tube and emerges from the spray chamber into the sample injector of the plasma torch. Double-pass spray chambers come in a variety of shapes, sizes, and materials, and are generally considered the most rugged design for routine use. Figure 3.8 shows a Scott

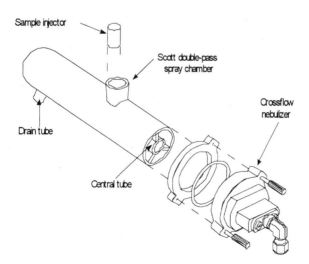

FIGURE 3.8 A Scott double pass spray chamber with crossflow nebulizer. (Courtesy of Perkin-Elmer Life and Analytical Sciences.)

double-pass spray chamber made from a polysulfide-type material, coupled to a crossflow nebulizer.

Cyclonic Spray Chamber

The cyclonic spray chamber operates by centrifugal force. Droplets are discriminated according to their size by means of a vortex produced by the tangential flow of the sample aerosol and argon gas inside the chamber. Smaller droplets are carried with the gas stream into the ICP-MS, while the larger droplets impinge on the walls and fall out through the drain. It is generally accepted that a cyclonic spray chamber has a higher sampling efficiency, which for clean samples, translate into higher sensitivity and lower detection limits. However, the droplet size distribution appears to be different from a double pass design, and for certain types of samples can give slightly inferior precision. Beres and coworkers [7] published a very useful study of the capabilities of a cyclonic spray chamber in 1994. Figure 3.9 shows a cyclonic spray chamber connected to a concentric nebulizer.

There are many other nonstandard sample introduction devices such as ultrasonic nebulization, membrane desolvation, high-efficiency nebulization, flow injection, direct injection, electrothermal vaporization, and laser ablation, which will not be described in this chapter. However, because they are becoming increasingly important, particularly as ICP-MS users are de-

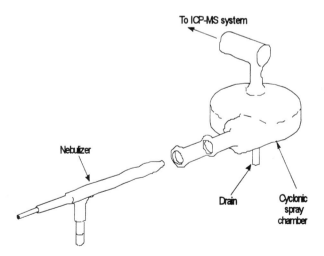

FIGURE 3.9 A cyclonic spray chamber (shown with concentric nebulizer). (From Ref. 7.)

manding higher performance and more flexibility, they will be covered in greater detail in Chapter 17 on "Alternate Sampling Accessories."

FURTHER READING

1. Browner RA, Boorn AW. Analytical Chemistry 1984; 56:786A–798A.
2. Sharp BL. Analytical Atomic Spectrometry 1980; 3:613.
3. Bates LC, Olesik JW. Journal of Analytical Atomic Spectrometry 1990; 5(3):239.
4. Houk RS. Analytical Chemistry 1986; 56:97.
5. Williams JG, Gray AL, Norman P, Ebdon L. Journal of Analytical Atomic Spectrometry 1987; 2:469–472.
6. Debrah E, Beres SA, Gluodennis TJ, Thomas RJ, Denoyer ER. Atomic Spectroscopy 1995; 16(7):197–202.
7. Beres SA, Bruckner PH, Denoyer ER. Atomic Spectroscopy 1994; 15(2):96–99.

4

Plasma Source

This chapter takes a look at the area where the ions are generated—the plasma discharge. It will give a brief historical perspective of some of the common analytical plasmas used over the years, and discusses the components that are used to create the inductively coupled plasma (ICP). It will then explain the fundamental principles of formation of a plasma discharge and how it is used to convert the sample aerosol into a stream of positively charged ions of low kinetic energy required by the ion focusing system and the mass spectrometer.

Inductively coupled plasmas (ICPs) are by far the most common type of plasma sources used in today's commercial ICP–optical emission spectrometry (OES) and ICP–mass spectrometry (MS) instrumentation. However, it was not always that way. In the early days, when researchers were attempting to find the ideal plasma source to use for spectrometric studies, it was not clear which approach would prove to be the most successful. In addition to inductively coupled plasmas, some of the other novel plasma sources developed were direct current plasmas (DCPs) and microwave-induced plasmas (MIPs). A DCP is formed when a gas (usually argon) is introduced into a high current flowing between 2 or 3 electrodes. Ionization of the gas produces a Y-shaped plasma. Unfortunately, early DCP instrumentation was prone to interference effects and had some usability and reliability problems. For these reasons, the technique never became widely accepted by the analytical community (1). However, its one major benefit was that it could aspirate high dissolved and/or suspended solids because there was no restrictive sample injector for the solid material to block. This feature alone made it very attractive for some laboratories and once the initial limitations of DCPs were better understood, the technique became more accepted. In fact, if you want a DCP excitation source coupled to an optical emission instrument today, an Echelle-based grating using a solid-state detector is commercially available (2).

Limitations in the DCP approach led to the development of electrodeless plasma of which the MIP was the simplest form. In this system, microwave energy (typically 100–200 W) is supplied to the plasma gas from an excitation cavity around a glass/quartz tube. The plasma discharge in the form of a ring is generated inside the tube. Unfortunately, even though the discharge achieves a very high power density, the high excitation temperatures exist only along a central filament. The bulk of the MIP never gets above 2000–3000 K, which means it was prone to very severe matrix effects. In addition, they were easily extinguished when aspirating liquid samples. For these reasons, they had limited success as an emission source because they were not considered robust enough for the analysis of real-world solution-based samples. However, they have gained acceptance as an ion source for mass spectrometry (3) and also as emission-based detectors for gas chromatography.

Because of the limitations of the DCP and MIP approaches, ICPs became the dominant area of research for both optical emission and mass spectrometric studies. As early as 1964, Greenfield and coworkers reported that an atmospheric pressure inductively coupled plasma coupled with optical emission spectrometry could be used for elemental analysis (4). Although crude by today's standards, it showed the enormous possibilities of the ICP as an excitation source and opened the door in the early 1980s to the even more exciting potential of using the ICP to generate ions (5).

THE PLASMA TORCH

Before we take a look at the fundamental principles behind the creation of an inductively coupled plasma used in ICP-MS, let us take a look at the basic components that are used to generate the source-a plasma torch, radio frequency (RF) coil and power supply. Figure 4.1 shows their proximity compared to the rest of the instrument, while Figure 4.2 is a more detailed view of the plasma torch and RF coil relative to the MS interface.

The plasma torch consists of three concentric tubes, which are normally made from quartz. In Figure 4.2, these are shown as the outer tube, middle tube, and sample injector. The torch can be either one piece, commonly known as the Fassel design where all three tubes are connected, or a demountable design where the tubes and the sample injector are separate. The gas (usually argon) that is used to form the plasma (plasma gas) is passed between the outer and middle tubes at a flow rate of ~12–17 L/min. A second gas flow (auxiliary gas) passes between the middle tube and the sample injector at ~1 L/min and is used to change the position of the base of the plasma relative to the tube and the injector. A third gas flow (nebulizer gas) also at ~1 L/min brings the sample, in the form of a fine droplet aerosol,

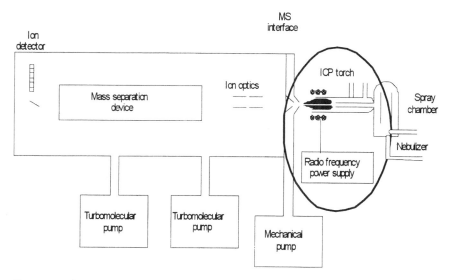

FIGURE 4.1   ICP-MS showing location of the plasma torch and RF power supply.

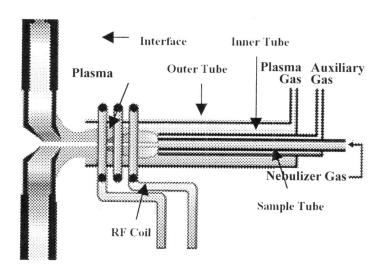

FIGURE 4.2   Detailed view of plasma torch and RF coil relative to the ICP-MS interface.

from the sample introduction system (for details refer to Chapter 3 on "Sample Introduction") and physically punches a channel through the center of the plasma. The sample injector is often made from other materials besides quartz, such as alumina, platinum, and sapphire, if highly corrosive materials need to be analyzed. Note that although argon is the most suitable gas to use for all three flows, there are analytical benefits in using other gases mixtures, especially in the nebulizer flow (6). The plasma torch is mounted horizontally and positioned centrally in the RF coil, approximately 10–20 mm from the interface. This can be seen in Figure 4.3, which shows a photograph of a plasma torch mounted in an instrument.

It must be emphasized that the coil used in an ICP-MS plasma is slightly different from the one used in ICP-OES, because in a plasma discharge, there is a potential difference of a few hundred volts produced by capacitive coupling between the RF coil and the plasma. In an ICP mass spectrometer, this would result in a secondary discharge between the plasma and the interface cone, which can negatively affect the performance of the instrument. To compensate for this, the coil must be grounded to keep the interface region as close to zero potential as possible. The full implications of this will be discussed in greater detail in Chapter 5 on the "Interface Region."

FIGURE 4.3 Photograph of a plasma torch mounted in an instrument. (Courtesy of Varian, Inc.)

FORMATION OF AN INDUCTIVELY COUPLED PLASMA DISCHARGE

Let us now discuss in greater detail the mechanism of formation of the plasma discharge. First, a tangential (spiral) flow of argon gas is directed between the outer and middle tube of a quartz torch. A load coil (usually copper) surrounds the top end of the torch and is connected to an RF generator. When RF power (typically 750–1500 W, depending on the sample) is applied to the load coil, an alternating current oscillates within the coil at a rate corresponding to the frequency of the generator. In most ICP generators this frequency is either 27 or 40 MHz (commonly known as megahertz or million cycles per second). This RF oscillation of the current in the coil causes an intense electromagnetic field to be created in the area at the top of the torch. With argon gas flowing through the torch, a high-voltage spark is applied to the gas causing some electrons to be stripped from their argon atoms. These electrons, which are caught up and accelerated in the magnetic field, then collide

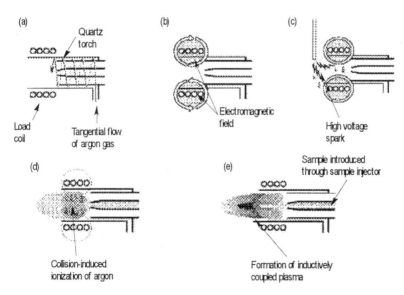

FIGURE 4.4 Schematic of an ICP torch and load coil showing how the ICP is formed. (a) A tangential flow of argon gas is passed between the outer and middle tube of the quartz torch. (b) RF power is applied to the load coil, producing an intense electromagnetic field. (c) A high-voltage spark produces free electrons. (d) Free electrons are accelerated by the RF field, causing collisions and ionization of the argon gas. (e) The ICP is formed at the open end of the quartz torch. The sample is introduced into the plasma via the sample injector (7).

with other argon atoms, stripping off still more electrons. This collision-induced ionization of the argon continues in a chain reaction, breaking down the gas into argon atoms, argon ions, and electrons, forming what is known as an ICP discharge. The ICP discharge is then sustained within the torch and load coil as RF energy is continually transferred to it through the inductive coupling process. The amount of energy required to generate argon ions in this process is approximately 15.8 eV (first ionization potential), which is enough energy to ionize the majority of the elements in the periodic table. The sample aerosol is then introduced into the plasma through a third tube called the sample injector. This whole process is conceptionally shown in Figure 4.4 (7).

THE FUNCTION OF THE RADIO FREQUENCY GENERATOR

Although the principles of an RF power supply have not changed since the work of Greenfield, the components have become significantly smaller. Some of the early generators that used nitrogen or air required 5–10 kW of power to sustain the plasma discharge-and literally took up half the room. Most of today's generators use solid-state electronic components, which means that vacuum power amplifier tubes are no longer required. This makes modern instruments significantly smaller, and, because vacuum tubes were notoriously unreliable and unstable, far more suitable for routine operation.

As mentioned previously, two frequencies have typically been used for ICP RF generators—27 and 40 MHz. These frequencies have been set aside specifically for RF applications of this kind, so they will not interfere with other communication-based frequencies. There has been much debate over the years as to which frequency gives the best performance (8,9). I think it is fair to say that although there have been several studies carried out, there does not appear to be any significant analytical advantage of one type over the other. In fact, of all the commercially available ICP-MS systems, there seems to be roughly an equal number of 27- and 40-MHz generators.

The more important consideration is the coupling efficiency of the RF generator to the coil. Most modern solid-state RF generators are about 70–75% efficient, which means that 70–75% of the delivered power actually makes it into the plasma. This was not always the case, and some of the older vacuum tube designed generators were notoriously inefficient—some of them experiencing over a 50% power loss. Another important criterion to consider is the way the matching network compensates for changes in impedance (a material's resistance to the flow of an electric current) produced by the sample's matrix components and/or differences in solvent volatility. In older-designed crystal-controlled generators, this was usually done with servo-driven capacitors. They worked very well with most sample types, but because they were mechanical devices, they struggled to compensate for very rapid impedance changes produced by some samples. As a result, it was fairly easy to extinguish the plasma, particularly when aspirating volatile organic solvents.

These problems were partially overcome by the use of free-running RF generators, where the matching network was based on electronic tuning of small changes in frequency brought about by the sample solvent and/or matrix components. The major benefit of this approach was that compensation for impedance changes was virtually instantaneous because there were no moving parts. This allowed for the successful analysis of many sample-types, which would most probably have extinguished the plasma of a crystal-controlled generator. However, because of improvements in electronic components over the years, the newer crystal-controlled generators appear to be equally as responsive as free-running designs.

IONIZATION OF THE SAMPLE

To better understand what happens to the sample on its journey through the plasma source, it is important to understand the different heating zones within the discharge. Figure 4.5 shows a cross-sectional representation of the discharge along with the approximate temperatures for different regions of the plasma.

As mentioned previously, the sample aerosol enters the injector via the spray chamber. When it exits the sample injector, it is moving at such a velocity that it physically punches a hole through the center of the plasma discharge. It then goes through a number of physical changes, starting at the

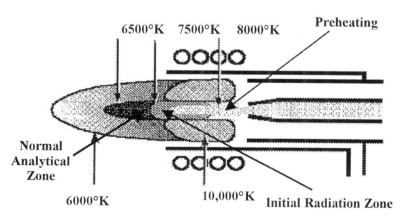

FIGURE 4.5   Different temperature zones in the plasma (7).

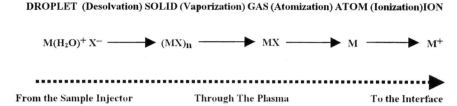

DROPLET (Desolvation) SOLID (Vaporization) GAS (Atomization) ATOM (Ionization)ION

$$M(H_2O)^+ X^- \longrightarrow (MX)_n \longrightarrow MX \longrightarrow M \longrightarrow M^+$$

From the Sample Injector   Through The Plasma   To the Interface

FIGURE 4.6 Mechanism of conversion of a droplet to a positive ion in the ICP.

preheating zone, continuing through the radiation zone before it eventually becomes a positively charged ion in the analytical zone. To explain this in a very simplistic way, let us assume that the element exists as a trace metal salt in solution. The fist step that takes place is desolvation of the droplet. With the water molecules stripped away, it then becomes a very small solid particle. As the sample moves further into the plasma, the solid particle changes first into a gaseous form and then into a ground state atom. The final process of conversion of an atom to an ion is achieved mainly by collisions of energetic argon electrons (and to a lesser extent, by argon ions) with the ground state atom (10). The ion then emerges from the plasma and is directed into the interface of the mass spectrometer (for details on the mechanisms of ion generation, please refer to Chapter 2 on "Principles of Ion Formation"). This process of conversion of droplets into ions is represented in Figure 4.6.

FURTHER READING

1. Gray AL. Analyst 1975; 100:289–299.
2. Coleman GN, Miller DE, Stark RW. Am Lab 1998; 30(4):33R.
3. Douglas DJ, French JB. Anal Chem 1981; 53:37–41.
4. Greenfield S, Jones IL, Berry CT. Analyst 1964; 89:713–720.
5. Houk RS, Fassel VA, Svec HJ. Dyn Mass Spectrom 1981; 6:234.
6. Lam JW, McLaren JW. J Anal At Spectrom 1990; 5:419–424.
7. Boss CB, Fredeen KJ. Concepts, Instrumentation and Techniques in Inductively Coupled Plasma Optical Emission Spectrometry. 2d ed. Norwalk, CT: Perkin Elmer Corporation, 1997.
8. Jarvis KE, Mason P, Platzner T, Williams JG. J Anal At Spectrom 1998; 13:689–696.
9. Vickers GH, Wilson DA, Hieftje GM. J Anal At Spectrom 1989; 4:749–754.
10. Hasegawa T, Haraguchi H. ICPs in Analytical Atomic Spectrometry. 2d ed. In: Montaser A, Golightly DW, ed. New York: VCH, 1992.

5

Interface Region

Chapter 5 takes a look at the interface region, which is probably the most critical area of the whole ICP-MS system. It gave the early pioneers of the technique the most problems to overcome. Although we take all the benefits of ICP-MS for granted, the process of taking a liquid sample, generating an aerosol that is suitable for ionization in the plasma and then sampling a representative number of analyte ions, transporting them through the interface, focusing them via the ion optics into the mass spectrometer, finally ending up with detection and conversion to an electronic signal, is not a trivial task. Each part of the journey has its own unique problems to overcome, but probably the most challenging is the movement of the ions from the plasma to the mass spectrometer.

The role of the interface region, which is shown in Figure 5.1, is to transport the ions efficiently, consistently and with electrical integrity from the plasma, which is at atmospheric pressure (760 Torr) to the mass spectrometer analyzer region at approximately 10^{-6} Torr.

This is first achieved by directing the ions into the interface region. The interface consists of two metallic cones with very small orifices, which are maintained at a vacuum of ~1–2 Torr with a mechanical roughing pump. After the ions are generated in the plasma, they pass into the first cone, known as the sampler cone, which has an orifice of 0.8–1.2 mm i.d. From there, they travel a short distance to the skimmer cone, which is generally smaller and more pointed than the sampler cone. The skimmer also has a much smaller orifice (typically 0.4–0.8 mm i.d.) than the sampler cone. Both cones are usually made of nickel but can be made of other materials such as platinum, which are far more tolerant to corrosive liquids. To reduce the effects of the high temperature plasma on the cones, the interface housing is water-cooled and made from a material that dissipates heat easily, such as copper or aluminum. The ions then emerge from the skimmer cone, where

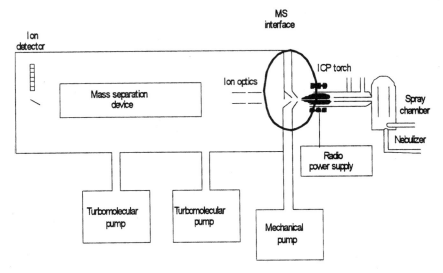

FIGURE 5.1   Schematic of ICP-MS, showing proximity of the interface region.

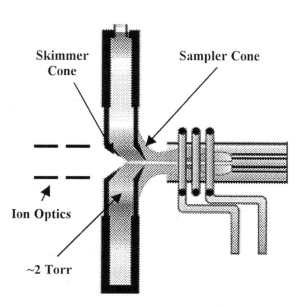

FIGURE 5.2   Detailed view of the interface region.

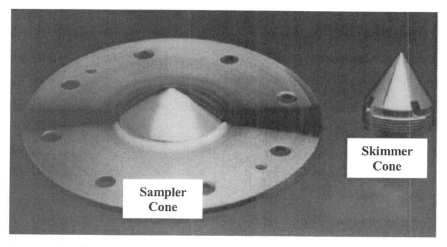

FIGURE 5.3   Close-up of the sampler and skimmer cones. (Courtesy of Varian, Inc.)

they are directed through the ion optics, and finally guided into the mass separation device. Figure 5.2 shows the interface region in greater detail, while Figure 5.3 shows a close up of the sampler and skimmer cone.

It should be noted that for most sample matrices, it is desirable to keep the total dissolved solids (TDS) below 0.2%, because of the possibility of deposition of the matrix components around the sampler cone orifice. This is not such a serious problem with short-term use but can lead to long-term signal instability if the instrument is being run for extended periods of time. The TDS levels can be higher (0.5–1%) when analyzing a matrix that forms a volatile oxide such as sodium chloride because once deposited on the cones, the volatile sodium oxide tends to revaporize without forming a significant layer that could potentially affect the flow through the cone orifice. In fact some researchers have reported running a 1:1 dilution of seawater (1.5% NaCl) for extended periods of time with good stability and no significant cone blockage—by careful optimization of the plasma RF power, sampling depth, and extraction lens voltage (1).

CAPACITIVE COUPLING

This process sounds fairly straight forward but proved to very problematic during the early development of ICP-MS, because of an undesired electrostatic (capacitive) coupling between the voltage on the load coil and the plasma discharge, producing a potential difference of 100–200 V. Although this potential is a physical characteristic of all inductively coupled plasma discharges, it was more serious in an ICP mass spectrometer, because the capacitive coupling created an electrical discharge between the plasma and the sampler cone. This discharge, commonly called the pinch effect or secondary discharge, shows itself as arcing in the region where the plasma is in contact with the sampler cone (2). This is seen very simplistically in Figure 5.4.

If not taken care of, this arcing can cause all kinds of problems, including an increase in doubly charged interfering species, a wide kinetic energy spread of sampled ions, formation of ions generated from the sampler cone, and decreased orifice lifetime. These were all problems reported by many of the early researchers into the technique (3,4). In fact, because the arcing increased with sampler cone orifice size, the source of the secondary discharge was originally thought to be the result of an electrogasdynamic effect, which produced an increase in electron density at the orifice (5). After many experiments, it was eventually realized that the secondary discharge was a result of electrostatic coupling of the load coil to the plasma. The problem was first eliminated by grounding the induction coil at the center, which had the effect of reducing the RF potential to a few volts. This can be seen in Figure 5.5 taken from one of the early papers, which shows the reduction in plasma potential as the coil is grounded at different positions (turns) along its length.

Originally, the grounding was achieved by attaching a physical grounding strap from the center turn of the coil to the interface housing. In today's instrumentation, the "grounding" is implemented in a number of

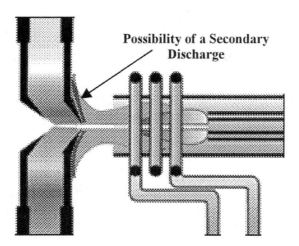

Possibility of a Secondary Discharge

FIGURE 5.4 Interface showing area affected by a secondary discharge.

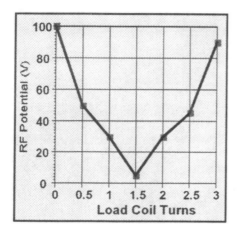

FIGURE 5.5 Reduction in plasma potential as the load coil is grounded at different positions (turns) along its length. (From Ref. 9.)

different ways, depending on the design of the interface. Some of the most popular designs include balancing the oscillator inside the circuitry of the RF generator (6), positioning a grounded shield or plate between the coil and the plasma torch (7), or by using two interlaced coil where the RF fields go in opposing directions (8). They all work differently but achieve a similar result of reducing or eliminating the secondary discharge.

ION KINETIC ENERGY

The impact of a secondary discharge cannot be overemphasized with respect to its effect on the kinetic energy of the ions being sampled. It is well documented that the energy spread of the ions entering the mass spectrometer must be as low as possible to ensure they can all be focused efficiently and with full electrical integrity by the ion optics and the mass separation device. When the ions emerge from the argon plasma, they will all have different kinetic energies, based on their mass-to-charge ratio. Their velocity should all be similar, because they are controlled by rapid expansion of the bulk plasma, which will be neutral, as long as it is maintained at zero potential. As the ion beam passes through the sampler cone into the skimmer cone, expansion will take place, but its composition and integrity will be maintained, assuming the plasma is neutral. This can be seen in Figure 5.6.

Electrodynamic forces do not play a role as the ions enter the sampler or the skimmer, because the distance over which the ions exert an influence

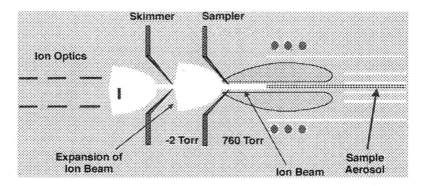

FIGURE 5.6 The composition of the ion beam is maintained as it passes through the interface, assuming a neutral plasma.

on each other (known as the Debye length) is small (typically 10^{-3}–10^{-4} mm) compared to the diameter of the orifice (0.5–1.0 mm) (9), as shown in Figure 5.7.

It is therefore clear that maintaining a neutral plasma is of paramount importance to guarantee electrical integrity of the ion beam as it passes through the interface region. If there is a secondary discharge present, it changes the electrical characteristics of the plasma, which will affect the

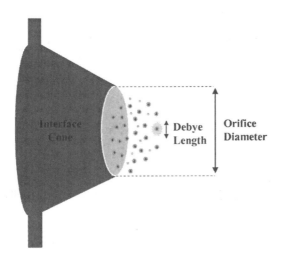

FIGURE 5.7 Electrodynamic forces do not affect the composition of the ion beam entering the sampler or the skimmer cone. (From Ref. 9.)

kinetic energy of the ions differently, depending on their mass-to-charge ratio. If the plasma is at zero potential, the ion energy spread is in the order of 5–10 eV. However, if there is a secondary discharge present, it results in a much wider spread of ion energies entering the mass spectrometer (typically 20–40 eV), which makes ion focusing far more complicated (9).

BENEFITS OF A WELL-DESIGNED INTERFACE

The benefits of a well-designed interface are not readily obvious if simple aqueous samples are being analyzed using only one set of operating conditions. However, it becomes more apparent when many different sample types are being analyzed, requiring different operating parameters. A true test of the design of the interface is when plasma conditions need to be changed, when the sample matrix changes, or when the ICP-MS is being used to analyze solid materials. Analytical scenarios like these have the potential to induce a secondary discharge, change the kinetic energy of the ions entering the mass spectrometer, and affect the tuning of the ion optics. It is therefore critical that the interface grounding mechanism can handle these types of real-world analytical situations, including:

- **Using cool-plasma conditions**: It is standard practice today to use cool plasma conditions (500–700 W power and 1.0–1.3 L/min nebulizer gas flow) to lower the plasma temperature and reduce argon-based polyatomic interferences such as $^{40}Ar^{16}O^+$, $^{40}Ar^+$, and $^{38}ArH^+$, in the determination of difficult elements such as $^{56}Fe^+$, $^{40}Ca^+$, and $^{39}K^+$. Such dramatic changes from normal operating conditions (1000 W, 0.8 L/min) will affect the electrical characteristics of the plasma.
- **Running organic solvents**: Analyzing oil or organic-based samples requires a chilled spray chamber (typically $-20\,°C$), or a membrane desolvation system to reduce the solvent-loading on the plasma. In addition, higher RF power ($\sim 1300–1500$ W) and lower nebulizer gas flow ($\sim 0.4–0.8$ L/min) is required to dissociate the organic components in the sample. A reduction in the amount of solvent entering the plasma combined with higher power and lower nebulizer gas flow translate into a hotter plasma and a change in its ionization mechanism.
- **Optimizing conditions for low oxides**: The formation of oxide species can be problematic in some sample types. For example, in geochemical applications it is quite common to sacrifice sensitivity, by lowering the nebulizer gas flow and increasing the RF power to reduce the formation of rare earth oxides—which can spectrally in-

terfere with the determination of other analytes. Unfortunately, these conditions will change the electrical characteristics of the plasma, which have the potential to induce a secondary discharge.

- **Using sampling accessories**: Sampling accessories such as membrane desolvators, laser ablation systems, and electrothermal vaporization devices are being used more routinely to enhance the flexibility of ICP-MS. The major difference between these sampling devices and a conventional liquid sample introduction system is they generate a "dry" sample aerosol, which requires totally different operating conditions compared to a conventional "wet" plasma. An aerosol that contains no solvent can have a dramatic affect on the ionization conditions in the plasma.

Although most modern ICP-MS interfaces have been designed to minimize the effects of the secondary discharge, it should not be taken for granted that they can all handle changes in operating conditions and matrix components with the same amount of ease. The most noticeable problems that have been reported include spectral peaks of the cone material appearing in the blank, erosion/discoloration of the sampling cones, widely different optimum plasma conditions (neb flow/RF power) for different masses, and frequent retuning of the ion optics (10,11). Chapter 20 on "How to Evaluate ICP-MS Instrumentation" goes into this subject in greater detail, but there is no question that the plasma discharge, interface region, and ion optics all have to be designed in concert to ensure the instrument can handle a wide range of operating conditions and sample types.

FURTHER READING

1. Plantz M, Elliott S. Application Note # ICP-MS 17. Varian Instruments, 1998.
2. Gray AL, Date AR. Analyst 1983; 108:1033.
3. Houk RS, Fassel VA, Svec HJ. Dyn Mass Spectrom 1981; 6:234.
4. Date AR, Gray AL. Analyst 1981; 106:1255.
5. Gray AL, Date AR. Dyn Mass Spectrom 1981; 6:252.
6. Tanner SD. J Anal At Spectrom 1995; 10:905.
7. Sakata K, Kawabata K. Spectrochim Acta 1994; 49B:1027.
8. Georgitus S, Plantz M. Winter Conference on Plasma Spectrochemistry, FP4, Fort Lauderdale, 1996.
9. Douglas DJ, French JB. Spectrochim Acta 1986; 41B(3):197.
10. Douglas DJ. Can J Spectrosc 1989; 34:2.
11. Fulford JE, Douglas DJ. Appl Spectrosc 1986; 40:7.

6

The Ion Focusing System

Chapter 6 takes a detailed look at the ion focusing system—a crucial area of the ICP-MS, where the ion beam is focused before it enters the mass analyzer. Sometimes known as the ion optics, it is composed of one or more ion lens components, which electrostatically steer the analyte ions in an axial (straight) or orthogonal (right-angled) direction from the interface region into the mass separation device. The strength of a well-designed ion focusing system is its ability to produce a flat signal response across the mass range, low background levels, good detection limits, and stable signals in real-world sample matrices.

Although the detection capability of ICP-MS is generally recognized as being superior to any of the other atomic spectroscopic techniques, it is probably most susceptible to the sample's matrix components. The inherent problem lies in the fact that ICP-MS is relatively inefficient—out of a million ions generated in the plasma, only one actually reaches the detector. One of the main contributing factors to the low efficiency is the higher concentration of matrix elements compared to the analyte, which has the effect of defocusing the ions and altering the transmission characteristics of the ion beam. This is sometimes referred to as a space charge effect and can be particularly severe when the matrix ions are of a heavier mass than the analyte ions (1). The roll of the ion focusing system is therefore to transport the maximum number of analyte ions from the interface region to the mass separation device, while rejecting as many of the matrix components and nonanalyte-based species as possible. Let us now discuss this process in greater detail.

ROLE OF THE ION OPTICS

The ion optics, which are shown in Figure 6.1, are positioned between the skimmer cone and mass separation device and consist of one or more electrostatically controlled lens components, maintained at a vacuum of approx-

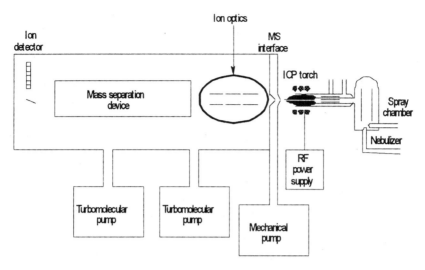

FIGURE 6.1   Position of ion optics relative to the plasma torch and interface region.

imately 10^{-3} Torr with a turbomolecular pump. They are not traditional optics that we associate with ICP emission or atomic absorption but made up of a series of metallic plates, barrels or cylinders, which have a voltage placed on them. The function of the ion optic system is to take ions from the hostile environment of the plasma at atmospheric pressure via the interface cones and steer them into the mass analyzer, which is under high vacuum. The nonionic species such as particulates, neutral species, and photons are prevented from reaching the detector either by using some kind of physical barrier, by positioning the mass analyzer off axis relative to the ion beam or by electro-statically bending the ions by 90° into the mass analyzer.

As mentioned in Chapters 4 and 5, the plasma discharge and interface region have to be designed in concert with the ion optics. It is absolutely critical that the composition and electrical integrity of the ion beam is maintained as it enters the ion optics. For this reason, it is essential that the plasma is at zero potential to ensure the magnitude and spread of ion energies is as low as possible (2).

A secondary, but also very important roll of the ion optic system, is to stop particulates, neutral species, and photons from getting through to the mass analyzer and the detector. These species cause signal instability and contribute to background levels, which ultimately affect the performance of the system. For example, if photons or neutral species reach the detector, they will elevate the noise of the background and therefore degrade detection capability. In addition, if particulates from the matrix penetrate further into the mass spectrometer region, they have the potential to deposit on lens components and in extreme cases get into the mass analyzer. In the short term this will cause signal instability and in the long term, increase the frequency of cleaning and routine maintenance.

There are basically three different approaches to reduce the chances of these undesirable species from making it into the mass spectrometer. The first method is to place a grounded metal stop (disk) behind the skimmer cone. This stop allows the ion beam to move around it but physically blocks the particulates, photons, and neutral species from traveling "downstream" (3). The second approach is to set the mass analyzer off axis to the ion lens system (in some systems, called a chicane design). The positively charged ions are then steered with the lens components into the mass analyzer, while the photons, neutral, and nonionic species are ejected out of the ion beam (4). The third and most recent development is to reflect the ion beam 90° with a "hollow" ion mirror (5). This allows the photons, neutrals, and solid particles

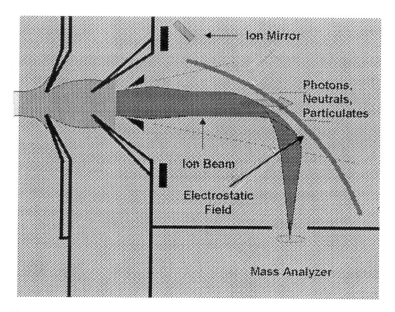

FIGURE 6.2   An ion focusing system which uses a hollow ion mirror to deflect the ion beam 90° to the mass analyzer, while allowing photons, neutrals, and solid particles to pass through. (Courtesy of Varian, Inc.)

to pass through, while the ions are reflected at right angles into an off-axis mass analyzer that incorporates curved fringe rod technology (6). The principle of this design is shown schematically in Figure 6.2.

It is also worth mentioning that some lens systems incorporate an extraction lens after the skimmer cone to electrostatically "pull" the ions from the interface region. This has the benefit of improving the transmission and detection limits of the low mass elements (which tend to be pushed out of the ion beam by the heavier elements), resulting in a more uniform response across the full mass range. In an attempt to reduce these space charge effects, some older designs have utilized lens components to accelerate the ions downstream. Unfortunately, this can have the effect of degrading the resolving power and abundance sensitivity (ability to differentiate an analyte peak from the wing of an interference) of the instrument, because of the much higher kinetic energy of the accelerated ions as they enter the mass analyzer (7).

DYNAMICS OF ION FLOW

To fully understand the roll of the ion optics in ICP-MS, it is important to get an appreciation of the dynamics of ion flow from the plasma through the interface region into the mass spectrometer. When the ions generated in the plasma emerge from the skimmer cone, there is a rapid expansion of the ion beam as the pressure is reduced from 760 Torr (atmospheric pressure) to approximately 10^{-3} to 10^{-4} Torr in the lens chamber with a turbomolecular pump. The composition of the ion beam immediately behind the cone is the same as the composition in front of the cone because the expansion at this stage is controlled by normal gas dynamics and not by electrodynamics. One of the main reasons for this is that in the ion sampling process, the Debye length (the distance over which ions exert influence on each other) is small compared to the orifice diameter of the sampler or skimmer cone. Consequently, there is little electrical interaction between the ion beam and the cone, and relatively little interaction between the individual ions in the beam. In this way, compositional integrity of the ion beam is maintained throughout the interface region (8). With this rapid drop in pressure in the lens chamber, electrons diffuse out of the ion beam. Because of the small size of the electrons relative to the positively charged ions, the electrons diffuse further from the beam than the ions, resulting in an ion beam with a net positive charge. This is represented schematically in Figure 6.3.

The generation of a positively charged ion beam is the first stage in the charge separation process. Unfortunately, the net positive charge of the ion beam means that there is now a natural tendency for the ions to repel each other. If nothing is done to compensate for this, ions of higher mass to charge

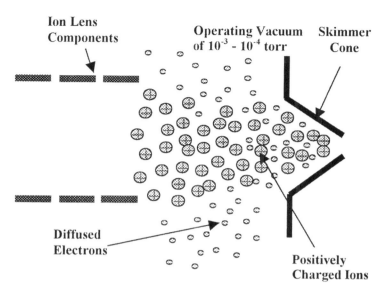

FIGURE 6.3 Extreme pressure-drop in the ion optic chamber produces diffusion of electrons, resulting in a positively charged ion beam.

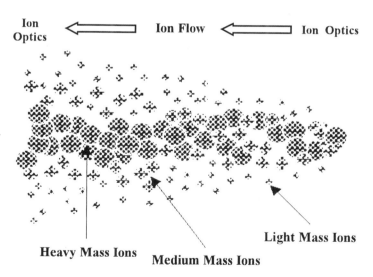

FIGURE 6.4 The degree of ion repulsion will depend on kinetic energy of the ions— the ones with high kinetic energy (heavy masses) will be transmitted in preference to ions with medium (medium masses) or low kinetic energy (light masses).

will dominate the center of the ion beam and force the lighter ions to the outside. The degree of loss will depend on the kinetic energy of the ions—the ones with high kinetic energy (high mass elements) will be transmitted in preference to ions with medium (mid mass elements) or low kinetic energy (low mass elements). This is shown in Figure 6.4. The second stage of charge separation is therefore to electrostatically steer the ions of interest back into the center of the ion beam, by placing voltages on one or more ion lens components. It should be emphasized that this is only possible if the interface is kept at zero potential, which ensures a neutral gas-dynamic flow through the interface, maintaining the compositional integrity of the ion beam. It also guarantees that the average ion energy and energy spread of each ion entering the lens systems are at levels optimum for mass separation. If the interface region is not grounded correctly, stray capacitance will generate a discharge between the plasma and sampler cone and increase the kinetic energy of the ion beam—making it very difficult to optimize the ion lens voltages (refer to Chapter 5 for details).

COMMERCIAL ION OPTIC DESIGNS

Over the years, there have been many different ion optic designs. Although they all have their own characteristics, they perform the same basic function to discriminate undesirable matrix or solvent-based ions, so that only the analyte ions are transmitted to the mass analyzer. The most common ion optics design used today consists of several lens components, which all have a specific role to play in the transmission of the analyte ions with the minimum of mass discrimination. With these multicomponent lens systems, the voltage can be optimized on every lens of the ion optics to achieve the desired ion specificity. This type of lens configuration has proved to be very durable over the years and shown to produce a uniform response across the mass range with very low background levels, particularly when combined with an off-axis mass analyzer (9). However, because of the interactive nature of parameters that affect the signal response, the more complex the lens system the more variables that have to be optimized. For this reason, if many different sample-types are being analyzed, extensive lens optimization procedures have to be carried out for each matrix or group of elements. This is not such a major problem, because the lens voltages are all computer-controlled and methods can be stored for every new sample scenario. However, it could be a factor if the instrument is being used for the routine analysis of many diverse sample types.

Another, more novel approach is to use just one cylinder lens, combined with a grounded stop—positioned just inside the skimmer cone as shown in Figure 6.5 (10).

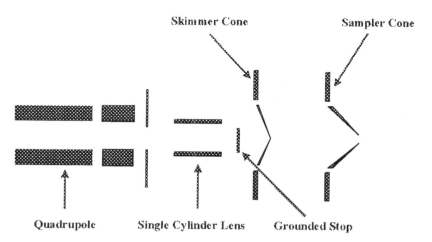

FIGURE 6.5 Schematic of a single ion lens and grounded stop system. (From Ref. 10.)

With this design, the voltage is dynamically ramped "on-the-fly," in concert with the mass scan of the analyzer (typically a quadrupole). The benefit of this approach is that the optimum lens voltage is placed on every mass in a multielement run to allow the maximum number of analyte ions through, while keeping the matrix ions down to an absolute minimum. This is represented in Figure 6.6, which shows a lens voltage scan of six elements Li, Co, Y, In, Pb, and U at 7, 59, 89, 115, 208, and 238 amu, respectively. It can be seen that each element has its own optimum value, which is then used to calibrate the system, so the lens can be ramp-scanned across the full mass range. This type of approach is typically used in conjunction with a grounded stop to act as a physical barrier to reduce particulates, neutral species, and photons from reaching the mass analyzer and detector. Although this design does not generate such a uniform mass response across the full range as an off-axis multilens system with an extraction lens, it appears to offer better long-term stability with real-world samples. It works well for many sample types but is most effective when low mass elements are being determined in the presence of high mass matrix elements.

Another approach is to use a simplified version of a collision cell to focus the ions into the mass analyzer. The collision cell in this mode is not used with a traditional collision gas but instead utilizes the multipole to act as an ion-focusing guide. This design of this type of ion lens system is usually incorporated with an off-axis quadrupole and a chicane-type deflector. The major advantage of this design is it gives extremely low background levels.

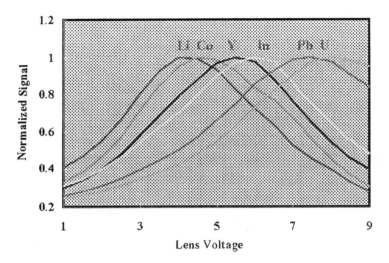

FIGURE 6.6 A calibration of optimum lens voltages is used to ramp scan the ion lens in concert with the mass scan of the analyzer. (From Ref. 10.)

A more recent development in ion focusing optics utilizes a parabolic electrostatic field created with an ion mirror to reflect and refocus the ion beam at 90° to the ion source (5). The ion mirror incorporates a hollow structure, which allows photons, neutrals, and solid particles to pass through it, while allowing ions to be reflected at right angles into the mass analyzer. The major benefit of this design is the very efficient way the ions are re-focused, offering the capability of extremely high sensitivity across the mass range, with very little sacrifice in oxide performance. In addition, there is very little contamination of the ion optics, because a vacuum pump sits behind the ion mirror to immediately remove these particles before they have a chance to penetrate further into the mass spectrometer. Removing these undesirable species and photons before they reach the detector, in addition to incorporating curved fringe rods prior to an off-axis mass analyzer, means that background levels are very low. Figure 6.7 shows a schematic of a quadru-pole-based ICP-MS that utilizes a 90° ion optic design (6,11).

It is also worth emphasizing that a number of ICP-MS systems offer what is called a high sensitivity option. These all work slightly differently but share similar components. By using a combination of slightly different cone geometry, higher vacuum at the interface, one or more extraction lens, and slightly modified ion optic design, they offer up to 10× the sensitivity of a traditional interface. However, in some systems, this increased sensitivity sometimes comes with slightly worse stability and an increase in background

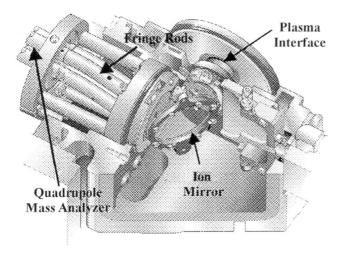

FIGURE 6.7 A 90° ion optic design used with curved fringe rods and an off-axis quadrupole mass analyzer. (Courtesy of Varian, Inc.)

levels, particularly for samples with a heavy matrix. To get around this, these kinds of samples typically need to be diluted before analysis—which has somewhat limited their applicability for real-world samples with high dissolved solids (13). However, they have found a use in nonliquid-based applications where high sensitivity is crucial—for example, in the analysis of small spots on the surface of a geological specimen using laser ablation ICP-MS. For this application, the instrument must offer high sensitivity, because a single laser pulse is often used to ablate very small amounts of the sample, which is then swept into the ICP-MS for analysis.

The importance of the ion focusing system cannot be overemphasized, because it has a direct bearing on the number of ions that find their way to the mass analyzer. As well as affecting background levels and instrument response across the entire mass range, it has a huge impact on both long- and short-term signal stability, especially in real-world samples. However, there are many different ways of achieving this. It is almost irrelevant whether the design of the ion optics is based on a dynamically scanned single ion lens or a multicomponent lens system; whether a grounded stop, an off-axis mass analyzer, or a right-angled bend is used to stop photons, particulates, and neutral species hitting the detector; or even whether an extraction lens is used. The most important consideration when evaluating an ion lens system is not the actual design but its ability to perform well with your sample matrices.

FURTHER READING

1. Olivares JA, Houk RS. Anal Chem 1986; 58:20.
2. Douglas DJ, French B. Spectrochim Acta 1986; 41B(3):197.
3. Tanner SD, Cousins LM, Douglas DJ. Appl Spectrosc 1994; 48:1367.
4. Potter D. Am Lab, vol. 26, July, 1994.
5. Kalinitchenko I. Ion Optical System for a Mass Spectrometer, Patent Number 750860, 1999.
6. Elliott S, Plantz M, Kalinitchenko L. Oral Paper 1360-8. Pittsburgh Conference, Orlando, FL, 2003.
7. Turner P. Paper at 2nd International Conference on Plasma Source Mass Spec. Durham, UK, 1990.
8. Tanner SD, Douglas DJ, French JB. Appl Spectrosc 1994; 48:1373.
9. Kishi Y. Agilent Technol Appl J August, 1997.
10. Denoyer ER, Jacques D, Debrah E, Tanner SD. At Spectrosc 1995; 16(1):1.
11. Kalinitchenko I. Mass Spectrometer Including a Quadrupole Mass Analyzer Arrangement, Patent applied for—WO 01/91159 A1.
12. Gibson BC. Paper at Surrey International Conference on ICP-MS, London, UK, 1994.

7

Mass Analyzers: Quadrupole Technology

The next four chapters deal with the heart of the system—the mass separation device. Sometimes called the mass analyzer, it is the region of the ICP-MS that separates the ions according to their mass-to-charge ratio. This selection process is achieved in a number of different ways, depending on the mass separation device, but they all have one common goal and that is to separate the ions of interest from all other nonanalyte, matrix, solvent, and argon-based ions. Quadrupole mass filters will be described in this chapter followed by magnetic sectors systems, time of flight mass spectrometers, and finally collision/reaction cell technology.

Although ICP-MS was commercialized in 1983, the first 10 years of its development was based on traditional quadrupole mass filter technology to separate the ions of interest. These worked exceptionally well for most applications but proved to have limitations when determining difficult elements or dealing with more complex sample matrices. This led to the development of alternative mass separation devices, which allowed ICP-MS to be used for applications which required higher resolution, faster data capture, and/or a reduction in polyatomic spectral interferences. Before we discuss these different mass spectrometers in greater detail, let us take a look at the proximity of mass analyzer in relation to the ion optics and the detector. Figure 7.1 shows this in greater detail.

    As can be seen, the mass analyzer is positioned between the ion optics and the detector and is maintained at a vacuum of approximately 10^{-6} Torr with an additional turbomolecular pump to the one that is used for the lens chamber. Assuming the ions are emerging from the ion optics at the optimum kinetic energy, they are ready to be separated according to their mass-to-charge ratio by the mass analyzer. There are basically four different kinds of commercially available mass analyzers—quadrupole mass filters, double

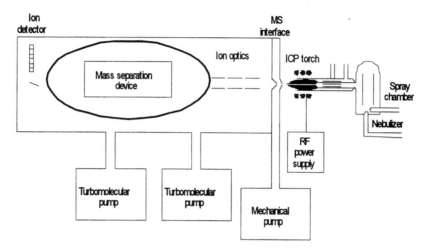

FIGURE 7.1 The mass separation device is positioned between the ion optics and the detector.

focusing magnetic sector, time of flight, and collision/reaction cell technology. They all have their own strengths and weaknesses, which will be discussed in greater detail in the next four chapters. Let us first begin with the most common of the mass separation devices used in ICP-MS—the quadrupole mass filter.

QUADRUPOLE MASS FILTER TECHNOLOGY

Developed in the early 1980s, quadrupole-based systems represent approximately 95% of all ICP-MS used today. This design was the first to be commercialized, and as a result, today's quadrupole ICP-MS technology is considered a very mature, routine, high-throughput trace element technique. A quadrupole consists of four cylindrical or hyperbolic metallic rods of the same length and diameter. They are typically made of stainless steel or molybdenum and sometimes coated with a ceramic coating for corrosion resistance. Quadrupoles used in ICP-MS are typically 15–20 cm in length, about 1 cm in diameter, and operate at a frequency of 2–3 MHz. Figure 7.2 shows a photograph of quadrupole system mounted in its housing.

BASIC PRINCIPLES OF OPERATION

A quadrupole operates by placing both a direct current (DC) field and a time-dependent alternating current (AC) of radio frequency on opposite pairs of

FIGURE 7.2 Photograph of a quadrupole system mounted in its housing. (Courtesy of Varian, Inc.)

the four rods. By selecting the optimum AC/DC ratio on each pair of rods, ions of a selected mass are then allowed to pass through the rods to the detector, while the others are unstable and ejected from the quadrupole. Figure 7.3 shows this in greater detail.

In this simplified example, the analyte ion (black) and four other ions (grey) have arrived at the entrance to the four rods of the quadrupole. When a

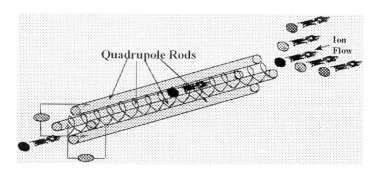

FIGURE 7.3 Schematic showing principles of mass separation using a quadrupole mass filter.

particular AC/DC potential is applied to the rods, the positive or negative bias on the rods will electrostatically steer the analyte ion of interest down the middle of the four rods to the end, where it will emerge and be converted to an electrical pulse by the detector. The other ions of different mass to charge will be unstable, pass through the spaces between the rods, and be ejected from the quadrupole. This scanning process is then repeated for another analyte at a completely different mass-to-charge ratio until all the analytes in a multiele-ment analysis have been measured. The process for the detection of one particular mass in a multielement run is represented in Figure 7.4. It shows a $^{63}Cu^+$ ion emerging from the quadrupole and being converted to an electrical pulse by the detector. As the AC/DC voltage of the quadrupole—correspond-ing to $^{63}Cu^+$—is repeatedly scanned, the ions as electrical pulses are stored and counted by a multichannel analyzer. This multichannel data acquisition system typically has 20 channels per mass and as the electrical pulses are counted in each channel, a profile of the mass is built up over the 20 channels,

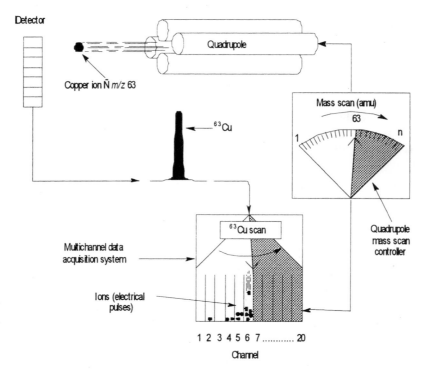

FIGURE 7.4 Profiles of different masses are built up using a multichannel data acquisition system. (Courtesy of PerkinElmer Life and Analytical Sciences.)

corresponding to the spectral peak of $^{63}Cu^+$. In a multielement run, repeated scans are made over the entire suite of analyte masses, as opposed to just one mass represented in this example.

Quadrupole scan rates are typically in the order of 2500 amu per second and can cover the entire mass range of 0–300 amu in about one-tenth of a second. However, real-world analysis speeds are much slower than this, and in practice, 25 elements can be determined in duplicate with good precision in 1–2 min, depending on the analytical requirements.

QUADRUPOLE PERFORMANCE CRITERIA

There are two very important performance specifications of a mass analyzer, which governs its ability to separate an analyte peak from a spectral interference. The first is resolving power (R), which in traditional mass spectrometry is represented by the equation: $R = m/\Delta m$, where m is the nominal mass at which the peak occurs and Δm is the mass difference between two resolved peaks (1). However, for quadrupole technology, the term resolution is more commonly used and is normally defined as the width of a peak at 10% of its height. The second specification is abundance sensitivity, which is the signal contribution of the tail of an adjacent peak at one mass lower and one mass higher than the analyte peak (2). Although they are somewhat related and both define the quality of a quadrupole, the abundance sensitivity is probably the most critical. If a quadrupole has good resolution, but poor abundance sensitivity, it will often prohibit the measurement of an ultra-trace analyte peak next to a major interfering mass.

Resolution

Let us now discuss this area in greater detail. The ability to separate different masses with a quadrupole is determined by a combination of factors including shape, diameter and length of the rods, frequency of quadrupole power supply, operating vacuum, applied RF/DC voltages, and the motion and kinetic energy of the ions entering and exiting the quadrupole. All these factors will have a direct impact on the stability of the ions as they travel down the middle of the rods and therefore the quadrupole's ability to separate ions of differing mass to charge. This is represented in Figure 7.5, which shows a simplified version of the Mathieu mass stability plot of two separate masses (A and B) entering the quadrupole at the same time (3).

Any of the RF/DC conditions shown under the left-hand peak (dark gray) will only allow mass A to pass through the quadrupole, while any combination of RF/DC voltages under the right-hand peak (light gray) plot will only allow mass B to pass through the quadrupole. If the slope of the

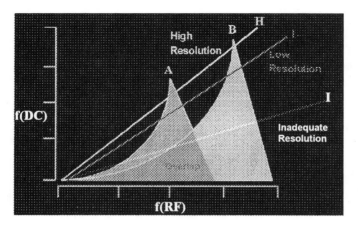

FIGURE 7.5 Simplified Mathieu stability diagram of a quadrupole mass filter, showing separation of two different masses—A and B. (From Ref. 3.)

RF/DC scan rate is steep, represented by the top line (high resolution), the spectral peaks will be narrow and masses A and B will be well separated. However, if the slope of the scan is shallow, represented by the middle line (low resolution), the spectral peaks will be wide and masses A and B will not be well separated. On the other hand, if the slope of the scan is too shallow, represented by the lower line (inadequate resolution), the peaks will overlap each other and the masses will pass through the quadrupole without being separated. In theory, the resolution of a quadrupole mass filter can be varied between 0.3 and 3.0 amu but is normally kept at 0.7–1.0 amu for most applications. However, improved resolution is always accompanied by a sacrifice in sensitivity as seen in Figure 7.6, which shows a comparison of the same mass at a resolution of 3.0, 1.0, and 0.3 amu.

It can be seen that the peak height at 3.0 amu is much larger than the peak height at 0.3 amu but, as expected, is also much wider. This would prohibit using a resolution of 3.0 amu with spectrally complex samples. Conversely, the peak width at 0.3 amu is very narrow, but the sensitivity is low. For this reason, a compromise between peak width and sensitivity normally has to be reached, depending on the application. This can clearly be seen in Figure 7.7, which shows a spectral overlay of two copper isotopes—$^{63}Cu^{+}$ and $^{65}Cu^{+}$—at resolution settings of 0.70 and 0.50 amu. In practice, the quadrupole is normally operated at a resolution of 0.7–1.0 amu, for the majority of applications.

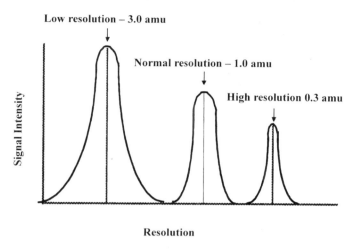

FIGURE 7.6 Sensitivity comparison of a quadrupole operated at 3.0-, 1.0-, and 0.3-amu resolution.

It is worth mentioning that most quadrupoles are operated in the first stability region, where resolving power is typically in the order of 500–600. If the quadrupole is operated in the second or third stability regions, resolving powers of 4000 (4) and 9000 (5), respectively, have been achieved. However, improving resolution using this approach has resulted in a significant loss of signal. Although there are ways of improving sensitivity, other problems have been encountered and as a result, to date, there are no commercial instruments available based on this design.

Some instruments can vary the peak width "on-the-fly," which means that the resolution can be changed between 3.0 and 0.3 amu for every analyte, in a multielement run. Although this appears to offer some benefits, in reality they are few and far between, and for the vast majority of applications, it is adequate to use the same resolution setting for every analyte. So, although quadrupoles can be operated at higher resolution (in the first stability region), up to now the slight improvement has not shown to be a practical benefit for most routine applications.

Abundance Sensitivity

It can be seen in Figure 7.7 that the tail of the spectral peaks drop-off more rapidly at the high mass end of the peak compared to the low mass end. The overall peak shape, particularly its low mass and high mass tail, is determined by the abundance sensitivity of the quadrupole, which is impacted by a

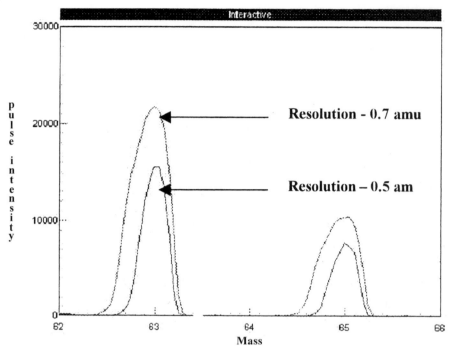

FIGURE 7.7 Sensitivity comparison of two copper isotopes—$^{63}Cu^+$ and $^{65}Cu^+$—at resolution settings of 0.70 amu and 0.50 amu.

combination of factors including design of the rods, frequency of the power supply, and operating vacuum (6). Although they are all important, probably the biggest impact on abundance sensitivity is the motion and kinetic energy of the ions as they enter and exit the quadrupole. If one looks at the Mathieu stability plot in Figure 7.5, it can be seen that the stability boundaries of each mass are less defined (not so sharp) on the low mass side than they are on the high mass side (3). As a result the characteristic of ion motion at the low mass boundary is different from the high mass boundary and is therefore reflected in poorer abundance sensitivity at the low mass side compared to the high mass side. The velocity and therefore the kinetic energy of the ions entering the quadrupole will affect the ion motion and as a result will have a direct impact on the abundance sensitivity. For that reason, factors that affect the kinetic energy of the ions, such as high plasma potential and the use of lens to accelerate the ion beam, will have a negative affect on the instrument's abundance sensitivity (7).

These are the fundamental reasons why the peak shape is not symmetrical with a quadrupole and explains why there is always a pronounced shoulder at the low mass side of the peak compared to the high mass side—as represented in Figure 7.8, which shows the theoretical peak shape of a nominal mass M. It can be seen that the shape of the peak at one mass lower $(M-1)$ is slightly different from the other side of the peak at one mass higher $(M+1)$ than the mass M. For this reason, the abundance sensitivity specification for all quadrupoles is always worse on the low mass side than the high mass side and is typically 1×10^{-6} at $M-1$ and 1×10^{-7} at $M+1$. In other words, an interfering peak of 1 million counts per second (cps) at $M-1$ would produce a background of 1 cps at M, while it would take an interference of 10 million cps at $M+1$ to produce a background of 1 cps at M.

Benefit of Good Abundance Sensitivity

An example of the importance of abundance sensitivity is shown in Figure 7.9. Figure 7.9A is a spectral scan of 50 ppm of the doubly charged europium

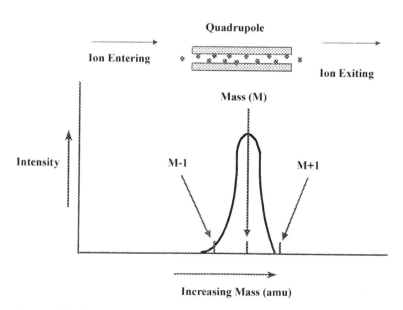

FIGURE 7.8 Ions entering the quadrupole are slowed down by the filtering process and produce peaks with a pronounced tail or shoulder at the low-mass end $(M - 1)$ compared to the high-mass end $(M + 1)$.

Doubly-charged ^{151}Eu^{++} (75.5 amu)

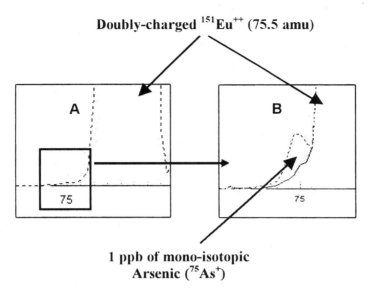

**1 ppb of mono-isotopic
Arsenic (^{75}As^{+})**

FIGURE 7.9 A low abundance sensitivity specification is critical to minimize spectral interferences, as shown by **A**, which represents a spectral scan of 50 ppm of ^{151}Eu^{2+} at 75.5 amu, and **B**, which clearly shows how the tail of the ^{151}Eu^{2+} elevates the spectral background of 1 ppb of As at mass 75. (Courtesy of Perkin-Elmer Life and Analytical Sciences.)

ion—^{151}Eu^{2+} at 75.5 amu (a doubly-charged ion is one with two positive charges, as opposed to a normal singly charged positive ion, and exhibits an m/z peak at half its mass). It can be seen that the intensity of the peak is so great that its tail overlaps the adjacent mass at 75 amu, which is the only available mass for the determination of arsenic. This is highlighted in Figure 7.9B, which shows an expanded view of the tail of the ^{151}Eu^{2+}, together with a scan of 1 ppb of As at mass 75. It can be seen very clearly that the ^{75}As^{+} signal lies on the sloping tail of the ^{151}Eu^{2+} peak. Measurement on a sloping background like this would result in a significant degradation in the arsenic detection limit, particularly as the element is mono-isotopic and no alternative mass is available. In this particular example a slightly higher resolution setting was also used (0.5 amu instead of 0.7 amu) to enhance the separation of the arsenic peak from the europium peak but nevertheless still emphasizes the importance of good abundance sensitivity in ICP-MS.

There are many different designs of quadrupole used in ICP-MS, all made from different materials with varied dimensions, shape, and physical characteristics. In addition, they are all maintained at a slightly different vacuum chamber pressure and operate at different frequencies. Theory tells us that hyperbolic rods should generate a better hyperbolic (elliptical) field than cylindrical rods—resulting in higher transmission of ions at higher resolution. It also tells us that a higher operating frequency means a higher rate of oscillation—and therefore separation—of the ions as they travel down the quadrupole. Finally, it is very well accepted that a higher vacuum produces less collisions between gas molecules and ions, resulting in a narrower spread in kinetic energy of the ions and therefore less of a tail at the low mass side of a peak. Given all these theoretical differences, in reality, the practical capabilities of most modern quadrupoles used in ICP-MS are very similar. However, there are some subtle differences in each instrument's measurement protocol and the software's approach to peak quantitation. This is such an important area that it will be discussed in greater detail in Chapter 12 on "Peak Measurement Protocol."

FURTHER READING

1. Adams F, Gijbels R, Van Grieken R. Inorganic Mass Spectrometry. New York: John Wiley and Sons, 1988.
2. Montasser A, ed. Inductively Coupled Plasma Mass Spectrometry. Berlin: Wiley-VCH, 1998.
3. Dawson PH, ed. Quadrupole Mass Spectrometry and its Applications. Amsterdam: Elsevier, 1976. reissued by AIP Press, Woodbury NY, 1995.
4. Du Z, Olney TN, Douglas DJ. J Am Soc Mass Spectrom 1997; 8:1230–1236.
5. Dawson PH, Binqi Y. Int J Mass Spectrom Ion Proc 1984; 56:25.
6. Potter D. Agilent Technol Appl Note 228–349, January, 1996.
7. Denoyer ER, Jacques D, Debrah E, Tanner SD. At Spectrosc 1995; 16(1):1.

8

Mass Analyzers: Double-Focusing Magnetic Sector Technology

Although quadrupole mass analyzers represent over 90% of all ICP-MS systems installed worldwide, limitations in their resolving power has led to the development of high-resolution spectrometers based on the double-focusing magnetic sector design. In this chapter we will take a detailed look at this very powerful mass separation device, which has found its niche in solving challenging application problems that require excellent detection capability, exceptional resolving power, and very high precision.

As discussed in Chapter 7, a quadrupole-based ICP-MS system typically offers a resolution of 0.7–1.0 amu. This is quite adequate for most routine applications, but has proved to be inadequate for many elements that are prone to argon-, solvent-, and/or sample-based spectral interferences. These limitations in quadrupoles drove researchers in the direction of traditional high-resolution, magnetic sector technology to improve quantitation by resolving the analyte mass away from the spectral interference (1). These ICP-MS instruments, which were first commercialized in the late 1980s, offered resolving power of up 10,000, compared to a quadrupole, which was on the order of ~300. This dramatic improvement in resolving power allowed difficult elements such as Fe, K, As, V, and Cr to be determined with relative ease, even in complex sample matrices.

MAGNETIC SECTOR MASS SPECTROMETRY: A HISTORICAL PERSPECTIVE

Mass spectrometers, using separation based on velocity focusing (2,3) and magnetic deflection (4,5), were first developed over 80 years ago, primarily to investigate isotopic abundances and calculate atomic weights. Although these designs were combined into one instrument in the 1930s to improve both sensitivity and resolving power (6,7), they were still considered rather bulky and expensive to build. For that reason, in the late 1930s and 1940s, magnetic field technology, and in particular the small-radius, sector design of Nier (8), became the preferred method of mass separation. Because Nier was a physicist, most of the early work performed with this design was used for isotope studies in the disciplines of earth and planetary sciences. However, it was the oil industry that accelerated the commercialization of mass spectrometry, because of its demand for fast and reliable analysis of complex hydrocarbons in oil refineries.

Once scanning magnetic sector technology became the most accepted approach of high-resolution mass separation in the 1940s, the challenges that lay ahead for mass spectroscopists were in the design of the ionization source—especially as the technique was being used more and more for the analysis of solids. The gas discharge ion source, which was developed for gases and high-vapor-pressure liquids, proved to be inadequate for most solid materials. For this reason, one of the first successful methods of ionizing solids was performed using the hot anode method (9), where the previously dissolved material was deposited on to a strip of platinum foil and evaporated by passing an electric current through it. Unfortunately, although there were variations of this approach that all worked reasonably well, the main drawback of a thermal evaporation technique is selective ionization. In other words, because of the different volatilities of the elements, it could not be guaranteed that the ion beam properly represented the compositional integrity of the sample.

It was finally the work done by Dempster in 1946 (10), using a vacuum spark discharge, based on a high-frequency, high-voltage spark, that led researchers to believe that it could be applied to sample electrodes and used as a general-purpose source for the analysis of solids. The breakthrough came in 1954 with the development of the first modern spark source mass spectrometer (SSMS) based on the Mattauch-Herzog mass spectrometer design (11). Using this design, Hannay and Ahearn, showed that it was possible to determine sub-ppm impurity levels directly in a solid material (12). Over the years, because of a demand for more stable ionization sources, lower detection capability, and higher precision, researchers were led in the direction of other techniques such as secondary ion mass spectrometry (SIMS) (13), ion microprobe mass spectrometry (IMMS) (14), and laser-induced mass spectrometry (LIMS) (15). Although they are considered somewhat complimentary to SSMS, they all had their own strengths and weaknesses depending on the analytical objectives for the solid material being analyzed. However, it should be emphasized that these techniques were predominantly used for microanalysis because only a very small area of the sample is vaporized. This meant that it could provide meaningful analytical data of the bulk material only if the sample was sufficiently homogeneous. For that reason, other ionization sources, which sampled a much larger area, such as the glow discharge, became much more practical for the bulk analysis of solids by mass spectrometry (16).

USE OF MAGNETIC SECTOR TECHNOLOGY FOR ICP-MS

Even though magnetic sector technology was the most common mass separation device for the analysis of inorganic compounds using traditional ion sources, it lost out to quadrupole technology when ICP-MS was first developed in the early 1980s. However, it was not until the mid-late 1980s, when the analytical community realized that quadrupole ICP-MS had serious limitations in its ability to resolve troublesome polyatomic spectral interferences, that researchers began to look at double-focusing magnetic sector technology to eliminate these kinds of problems. Initially it was found to be unsuitable as a separation device for an ICP because of the high voltage required to accelerate the ions into the mass analyzer. This high potential at

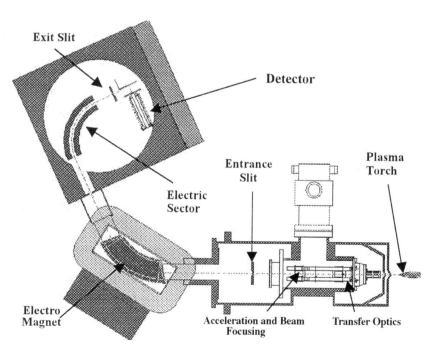

FIGURE 8.1 Schematic of a reverse Nier–Johnson double-focusing magnetic sector mass spectrometer. (From Ref. 17.)

the interface region dramatically changed the energy of the ions entering the mass spectrometer and therefore made it very difficult to steer the ions through the ion optics and still maintain a narrow spread of ion kinetic energies. For this reason, basic changes had to be made to the ion acceleration mechanism in order for magnetic sector technology to be successfully used as a separation device for ICP-MS. This was a significant challenge when magnetic sector systems were first developed in the late 1980s. However, by the early 1990s, one instrument manufacturer in particular solved this problem by moving the high-voltage components away from the plasma and interface closer to the mass spectrometer. Modern instrumentation has typically been based on two different approaches-the "standard" and "reverse" Nier–Johnson geometry. Both these designs, which use the same basic principles, consist of two analyzers—a traditional electromagnet and an electrostatic analyzer (ESA). In the standard (sometimes called forward) design, the ESA is positioned before the magnet, and in the reverse design it is positioned after the magnet. A schematic of the reverse Nier–Johnson spectrometer is shown in Figure 8.1 (17).

PRINCIPLES OF OPERATION OF MAGNETIC SECTOR SYSTEMS

The original concept of magnetic sector technology was to scan over a large mass range by varying the magnetic field over time with a fixed acceleration voltage. During a small window in time, which was dependant on the resolution chosen, ions of a particular mass to charge are swept passed the exit slit to produce the characteristic flat top peaks. As the resolution of a magnetic sector instrument is independent of mass, ion signals, particularly at low mass, are far apart. Unfortunately this results in a relatively long time being spent scanning and settling the magnet. This was not such a major problem for qualitative analysis or mass spectral fingerprinting of unknown compounds, but proved to be impractical for rapid trace element analysis, where you had to scan to individual masses, slow down, settle the magnet, stop, take measurements, and then scan to the next mass. However, by using the double-focusing approach, the ions are sampled from the plasma in a conventional manner and then accelerated in the ion optic region to a few kilovolts (kV) before they enter the mass analyzer. The magnetic field, which is dispersive with respect to ion energy and mass, then focuses all the ions with diverging angles of motion from the entrance slit. The ESA, which is dispersive only with respect to ion energy, then focuses all the ions onto the exit slit, where the detector is positioned. If the energy dispersion of the magnet and ESA are equal in magnitude but opposite in direction, they will focus both ion angles (first focusing) and ion energies (second or double focusing) when combined together. Changing the electrical field in the opposite direction during the cycle time of the magnet (in terms of the mass passing the exit slit) has the effect of "freezing" the mass for detection. Then as soon as a certain magnetic field strength is passed, the electric field is set to its original value and the next mass is "frozen." The voltage is varied on a per mass basis, allowing the operator to scan only the mass peaks of interest rather than the full mass range (18,19).

Note that although this approach represents an enormous time saving over traditional magnet scanning technology, it is still slower than quadrupole-based instruments. The inherent problem lies in the fact that a quadrupole can be electronically scanned faster than a magnet. Typical speeds for a full mass scan (0–250 amu) of a magnet are about, approximately 200 msec compared to 100 msec for a quadrupole. In addition, it takes much longer for a magnet to slow down, settle, and stop to take measurements—typically 20 msec compared to 1–2 msec for a quadrupole. So, although in practice, the electric scan dramatically reduces the overall analysis time, modern double-focusing magnetic sector ICP-MS systems are still slower than state-of-the-art quadrupole instruments, which makes them less than ideal for rapid, high-throughput multielement applications.

Resolving Power

As mentioned previously, most commercial magnetic sector ICP-MS systems offer up to 10,000 resolving power (5% peak height/10% valley definition), which is high enough to resolve most spectral interferences. It is worth emphasizing that resolving power (R), is represented by the equation: $R = m/\Delta m$, where m is the nominal mass at which the peak occurs and Δm is the mass difference between two resolved peaks (20). In a quadrupole, the resolution is selected by changing the ratio of the RF/DC voltages on the quadrupole rods. However, because a double-focusing magnetic sector instrument involves focusing ion angles and ion energies, mass resolution is achieved by using two mechanical slits—one at the entrance to the mass spectrometer and another at the exit, prior to the detector. Varying resolution is achieved by scanning the magnetic field under different entrance and exit slit width conditions. Similar to optical systems, low resolution is achieved by using wide slits, whereas high resolution is achieved with narrow slits. Varying the width of both the entrance and exit slits effectively changes the operating resolution.

However, it should be emphasized that similar to optical spectrometry, as the resolution is increased, the transmission decreases. So even though extremely high resolution is available, detection limits will be compromised under these conditions. This can be seen in Figure 8.2, which shows a plot of resolution against ion transmission. It can be seen that a resolving power of 400 produces 100% transmission, but at a resolving power of 10,000, only ~2% is achievable. This dramatic loss in sensitivity could be an issue if low

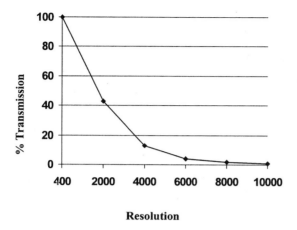

Resolution

FIGURE 8.2   Ion transmission with a magnetic sector instrument decreases as the resolution increases.

detection limits are required in spectrally complex samples that require the highest possible resolution. However, spectral demands of this nature are not very common. Table 8.1 shows the resolution required to resolve fairly common polyatomic interferences from a selected group of elemental isotopes, together with the achievable ion transmission.

Figure 8.3 is a comparison between a quadrupole and a magnetic sector instrument of one of the most common polyatomic interferences—$^{40}Ar^{16}O^+$ on $^{56}Fe^+$, which requires a resolution of 2504 to separate the peaks. Figure 8.3a shows a spectral scan of $^{56}Fe^+$ using a quadrupole instrument. What it does not show is the massive polyatomic interference $^{40}Ar^{16}O^+$ (produced by oxygen ions from the water combining with argon ions from the plasma) completely overlapping the $^{56}Fe^+$. It shows clearly that these two masses are unresolvable with a quadrupole. If that same spectral scan is performed on a magnetic sector-type instrument, the result is the scan shown in Figure 8.3b (21). To see the spectral scan on the same scale, it was necessary to examine a much smaller range. For this reason, 0.100 amu window was taken, as indicated by the dotted lines.

OTHER BENEFITS OF MAGNETIC SECTOR INSTRUMENTS

Besides high resolving power, another attractive feature of magnetic sector technology is its very high sensitivity combined with extremely low background levels. High ion transmission in low-resolution mode translates into sensitivity specifications of up to 1 billion counts per second (bcps) per ppm, while background levels resulting from extremely low dark current noise are

TABLE 8.1   Resolution Required to Resolve Some Common Polyatomic Interferences from a Selected Group of Isotopes

Isotope	Matrix	Interference	Resolution	Transmission (%)
$^{39}K^+$	H_2O	$^{38}ArH^+$	5,570	6
$^{40}Ca^+$	H_2O	$^{40}Ar^+$	199,800	0
$^{44}Ca^+$	HNO_3	$^{14}N^{14}N^{16}O^+$	970	80
$^{56}Fe^+$	H_2O	$^{40}Ar^{16}O^+$	2,504	18
$^{31}P^+$	H_2O	$^{15}N^{16}O^+$	1,460	53
$^{34}S^+$	H_2O	$^{16}O^{18}O^+$	1,300	65
$^{75}As^+$	HCl	$^{40}Ar^{35}Cl^+$	7,725	2
$^{51}V^+$	HCl	$^{35}Cl^{16}O^+$	2,572	18
$^{64}Zn^+$	H_2SO_4	$^{32}S^{16}O^{16}O^+$	1,950	42
$^{24}Mg^+$	Organics	$^{12}C^{12}C^+$	1,600	50
$^{52}Cr^+$	Organics	$^{40}Ar^{12}C^+$	2,370	20
$^{55}Mn^+$	HNO_3	$^{40}Ar^{15}N^+$	2,300	20

typically 0.1–0.2 cps. This compares to sensitivity of 10–50 mcps and background levels of ~10 cps for a quadrupole instrument. For this reason, detection limits, especially for high-mass elements such as uranium, where high resolution is generally not required, are typically 5 to 10 times better than a quadrupole-based instrument.

Besides good detection capability, another of the recognized benefits of the magnetic sector approach is its ability to quantitate with excellent precision. Measurement of the characteristically flat-topped spectral peaks translate directly into high precision data. As a result, in the low-resolution mode, relative standard deviation (RSD) values of 0.01–0.05% are fairly common, which makes them an ideal tool for carrying out high-precision isotope ratio work (22). Although precision is usually degraded as resolution is increased, modern instrumentation with high-speed electronics and low mass bias are still capable of precision values of <0.1% RSD in medium- or high-resolution mode (23).

The demand for ultra-high-precision data, particularly in geochemistry, has led to the development of instruments dedicated to isotope ratio analysis. These are based on the double-focusing magnetic sector design, but instead of using just one detector, these instruments use multiple detectors. Often referred to as multicollector systems, they can detect and measure multiple ion signals at exactly the same time. As a result of this simultaneous measurement approach, they are recognized as producing the ultimate in isotope ratio precision (24).

There is no question that double-focusing magnetic sector ICP-MS systems are no longer a novel analytical technique. They have proved them-

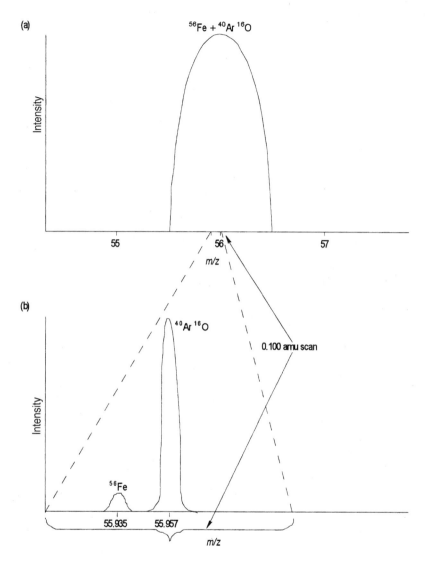

FIGURE 8.3 Comparison of resolution between a quadrupole (a) and a magnetic sector instrument (b) for the polyatomic interference of $^{40}Ar^{16}O^+$ on $^{56}Fe^+$. (From Ref. 21).

selves to be a valuable addition to the trace element toolkit, particularly for challenging applications that require good detection capability, exceptional resolving power, and/or very high precision. And although perhaps they are not competition for quadrupole instruments when it comes to rapid, high-sample-throughput applications, the scan speeds of modern systems have been improved considerably over the past few years. For that reason, they can now be considered a viable alternative to quadrupoles for carrying out multi-element determinations on transient peaks using laser ablation (25) or chromatographic separation devices (26).

FURTHER READING

1. Bradshaw N, Hall EFH, Sanderson NE. J Anal At Spectrom 1989; 4:801–803.
2. Aston FW. Philos Mag 1919; 38:707.
3. Costa JL. Ann Phys 1925; 4:425.
4. Dempster AJ. Phys Rev 1918; 11:316.
5. Swann WFG. J Franklin Inst 1930; 210:751.
6. Dempster AJ. Proc Am Philos Soc 1935; 75:755.
7. Bainbridge KT, Jordan EB. Phys Rev 1936; 50:282.
8. Nier AO. Rev Sci Instrum 1940; 11:252.
9. Thomson GP. Philos Mag 1921; 42:857.
10. Dempster AJ. MDDC 370. Washington, DC: U.S. Department of Commerce, 1946.
11. Mattauch J, Herzog R. Z Phys 1934; 89:786.
12. Hannay NB, Ahearn AJ. Anal Chem 1954; 26:1056.
13. Honig RE. J Appl Phys 1958; 29:549.
14. Castaing R, Slodzian G. J Microsc 1962; 1:395.
15. Honig RE, Wolston JR. Appl Phys Lett 1963; 2:138.
16. Coburn JW. Rev Sci Instrum 1970; 41:1219.
17. Compiled from poster presentation at 2nd Regensburg Symposium on Mass Spectrometry for Elemental Analysis, given by U Geismann, U Greb, Oct 5–8, 1993, and Finnigan MAT Element 2 Brochurem, 1998.
18. Hutton R, Walsh A, Milton D, Cantle J. ChemSA 1991; 17:213–215.
19. Geismann U, Greb U. Fresenius' J Anal Chem 1994; 350:186–193.
20. Adams F, Gijbels R, Van Grieken R. Inorganic Mass Spectrometry. New York: John Wiley and Sons, 1988.
21. Greb U, Rottman L. Labor Praxis, August 1994; 18:42–47.
22. Vanhaecke F, Moens L, Dams R, Taylor R. Anal Chem 1996; 68:567.
23. Hamester M, Wiederin D, Willis J, Keri W, Douthitt CB. Fresenius' J Anal Chem 1999; 364:495–497.
24. Walder J, Freeman PA. J Anal At Spectrom 1992; 7:571.
25. Shuttleworth S, Kremser D. J Anal At Spectrom 1998; 13:697–699.
26. Klueppel D, Jakubowski N, Messerschmidt J, Stuewer D, Klockow D. J Anal At Spectrom 1998; 13:255.

9

Mass Analyzers: Time of Flight Technology

Let us turn our attention to the most recent mass separation device to be commercialized—time of flight (TOF) technology. Although the first TOF mass spectrometer was first described in the literature in the late 1940s (1), it has taken over 50 years to adapt it for use with a commercial ICP mass spectrometer. The recent growth in TOF-ICP-MS sales has come about because of its unique ability to sample all ions generated in the plasma at exactly the same time, which is ideally suited for multielement determinations of rapid transient signals, high-precision ratio analysis, and rapid data acquisition.

BASIC PRINCIPLES OF TOF

The simultaneous nature of sampling ions in TOF offers distinct advantages over traditional scanning (sequential) quadrupole technology for ICP-MS applications where large amounts of data need to be captured in a short amount of time. To understand the benefits of this mass separation device, let us first take a look at its fundamental principles. All time-of-flight mass spectrometers are based on the same principle that the kinetic energy (KE) of an ion is directly proportional to its mass (m) and velocity (V). This can be represented by the equation:

$$KE = \frac{1}{2}mV^2$$

Therefore if a population of ions—all with different masses—is given the same KE by an accelerating voltage (U), the velocities of the ions will all be different, based on their masses. This principle is then used to separate ions

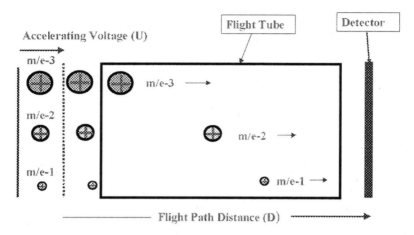

FIGURE 9.1   Principles of ion detection using time-of-flight technology, showing separation of three different masses in the time domain.

of different mass-to-charge (m/e) in the time (t) domain, over a fixed flight path distance (D), represented by the equation:

$$m/e = \frac{2Ut^2}{D^2}$$

This is schematically shown in Figure 9.1, which shows three ions of different mass-to-charge (m/e 1–3) being accelerated into a "flight tube" and arriving at the detector at different times. It can be seen that, based on their velocities, the lightest ion arrives first, followed by the medium mass ion, and finally the heaviest one. Using flight tubes of 1 m in length, even the heaviest ions typically take less than 50 μsec to reach the detector. This translates into approximately 20,000 mass spectra per second—3 orders of magnitude faster than the sequential scanning mode of a quadrupole system.

COMMERCIAL DESIGNS

Although this process sounds fairly straightforward, it is not a trivial task to sample the ions in a simultaneous manner from a continuous source of ions being generated in the plasma discharge. There are basically two different sampling approaches that are used in commercial TOF mass analyzers. They are the orthogonal design (2), where the flight tube is positioned right angles to the sampled ion beam, and the axial design (3), where the flight tube is in the same axis as the ion beam. In both designs, all ions that contribute to the mass spectrum are sampled through the interface cones, but instead of being focused into the mass filter in the conventional way, packets (groups) of ions are electrostatically injected into the flight tube at exactly the same time. With the orthogonal approach, an accelerating potential is applied at right angles to the continuous ion beam from the plasma source. The ion beam is then "chopped" by using a pulsed voltage supply coupled to the orthogonal accelerator to provide repetitive voltage "slices" at a frequency of a few kilohertz. The "sliced" packets of ions, which are typically tall and thin in cross section (in the vertical plane), are then allowed to "drift" into the flight tube where the ions are temporally resolved according to their differing velocities. This is shown schematically in Figure 9.2.

The axial approach is similar in design to the orthogonal approach, except that an accelerating potential is applied axially (in the same axis) to the incoming ion beam as it enters the extraction region. Because the ions are in the same plane as the detector, the beam has to be modulated using an electrode grid to repel the "gated" packet of ions into the flight tube. This kind of modulation generates an ion packet that is long and thin in cross section (in the horizontal plane). The different masses are then resolved in the time domain in a similar manner to the orthogonal design. An on-axis TOF system is schematically shown in Figure 9.3.

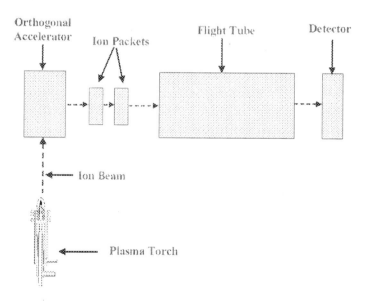

FIGURE 9.2 A schematic of an orthogonal acceleration TOF analyzer. (From Ref. 4.)

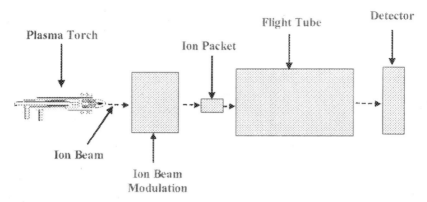

FIGURE 9.3 A schematic of an axial acceleration TOF analyzer. (From Ref. 4.)

Figures 9.2 and 9.3 represent a rather simplistic explanation of TOF principles of operation. In practice, there are many complex ion-focusing components that make up a TOF mass analyzer to ensure that the maximum number of analyte ions reaches the detector and that also undesired photons, neutral species, and interferences are ejected from the ion beam. Some of these components are seen in Figure 9.4, which shows a more detailed view of a commercial axial TOF system. The ions that pass through the interface are

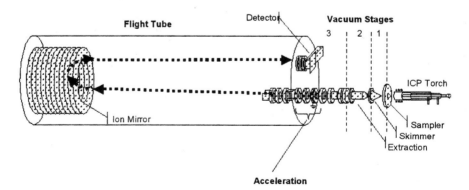

FIGURE 9.4 A more detailed view of a typical on-axis TOF analyzer, showing some of the ion steering components. (Courtesy of Leco Instruments.)

extracted and accelerated into the flight tube. The packets of extracted ions are then steered towards an ion mirror (or reflectron) and deflected back through 180°, where they are detected using a channel electron multiplier or discrete dynode detector. The reflectron in this design functions as an energy compensation device, so that different ions of the same mass arrive at the detector at the same time. Although the orthogonal design uses different components, the ion-steering principles are very similar.

DIFFERENCES BETWEEN ORTHOGONAL AND ON-AXIS TOF

Although there are real benefits of using TOF over quadrupole technology for some ICP-MS applications, there are also subtle differences in the capabilities of each type of TOF design. Without getting into the advantages and disadvantages of different commercial instrumentation, it is worth presenting the major differences between the orthogonal and on-axis approaches and comparing them with today's quadrupole-based instruments. Some of these differences include the following.

Sensitivity

The axial approach tends to produce higher ion transmission because the steering components are in the same plane as the ion generation system (plasma) and the detector. This means that the direction and the magnitude of greatest energy dispersion are along the axis of the flight tube. In addition, when ions are extracted orthogonally, the energy dispersion can produce angular divergence of the ion beam resulting in poor transmission efficiency. However, based on current evidence, the sensitivity of both TOF designs is generally an order of magnitude lower than the latest commercial quadrupole instruments.

Background Levels

The on-axis design tends to generate higher background levels because neutral species and photons stand a greater chance of reaching the detector. This results in background levels in the order of 20–50 cps—approximately an order of magnitude higher than the orthogonal design. However, because the ion beam in the axial design has a smaller cross section, a smaller detector can be used, which generally has better noise characteristics. In comparison, most commercial quadrupole instruments offer background levels of 1–10 cps depending on the design.

Duty Cycle

This is usually defined as the fraction (percentage) of extracted ions that actually make it into the mass analyzer. Unfortunately, with a TOF-ICP mass spectrometer that has to use "pulsed" ion packets from a continuous source of ions generated in the plasma, this process is relatively inefficient. It should also be emphasized that although the ions are sampled at the same time, detection is not simultaneous because of different masses arriving at the detector at different times. The difference between the sampling mechanisms of orthogonal and axial TOF designs translates into subtle differences in their duty cycles.

With the orthogonal design, duty cycle is defined by the width of the extracted ion packets, which are typically tall and thin in cross section (as shown in Figure 9.2). In comparison, the duty cycle of an axial design is defined by the length of the extracted ion packet, which is typically wide and thin in cross section (as shown in Figure 9.3). Duty cycle can be improved by changing the cross-sectional area of the ion packet, but, depending on the design, it is generally at the expense of resolution. However, this is not a major issue because TOF instruments are generally not used for high-resolution ICP-MS applications. In practice, the duty cycles for both orthogonal and axial designs are in the order of 15–20%.

Resolution

The resolution of the orthogonal approach is slightly better because of its two-stage extraction/acceleration mechanism. Because a pulse of voltage pushes the ions from the extraction area into the acceleration region, the major energy dispersion lies along the axis of ion generation. For this reason, the energy spread is relatively small in the direction of extraction compared to the axial approach, resulting in better resolution. However, the resolving power of commercial TOF-ICP-MS systems is typically in the order of 500–2000, depending on the mass region, which makes them inadequate to resolve many of the problematic polyatomic species encountered in ICP-MS. In comparison, commercial high-resolution systems based on the double-focusing magnetic sector design offer resolving power up to 10,000, while commercial quadrupoles achieve 300–400.

Mass Bias

This is also known as mass discrimination and is the degree to which ion transport efficiency varies with mass. All instruments show some degree of mass bias and are usually compensated for by measuring the difference between the theoretical and the observed ratio of two different isotopes of the same element. In TOF, the velocity (energy) of the initial ion beam will affect the instrument's mass bias characteristics. In theory, it should be less with the axial design because the extracted ion packets do not have any velocity in a direction perpendicular to the axis of the flight tube, which could potentially impact their transport efficiency.

BENEFITS OF TOF TECHNOLOGY FOR ICP-MS

It should be emphasized that these performance differences between the two designs are subtle and should not detract from the overall benefits of the TOF approach for ICP-MS. As mentioned earlier, a scanning device like a quadrupole can only detect one mass at a time, which means that there is always a compromise between number of elements, detection limits, precision, and overall measurement time. However, with the TOF approach, the ions are sampled at exactly the same moment in time, which means that multielement data can be collected with no significant deterioration in quality. The ability of a TOF system to capture a full mass spectrum, approximately 3 orders of magnitude faster than a quadrupole, translates into three major benefits—multielement determinations in a fast transient peak, improved precision, especially for isotope ratioing techniques, and rapid data acquisition. Let us look at these in greater detail.

Rapid Transient Peak Analysis

Probably, the most exciting potential for TOF-ICP-MS is in the multielement analysis of a rapid transient signal generated by sampling accessories like laser ablation (LA) (5), electrothermal vaporization (ETV) (6), and flow injection systems (7). Although a scanning quadrupole can be used for this type of analysis, it struggles to produce high-quality multielement data when the transient peak lasts only a few seconds. The simultaneous nature of TOF instrumentation makes it ideally suited for this type of analysis because the entire mass range can be collected in less than 50 μsec. In particular, when used with an ETV system, the high acquisition speed of TOF can help to reduce matrix-based spectral overlaps by resolving them from the analyte masses in the temperature domain (6). There is no question that TOF technology is ideally suited (probably more than any other design of ICP mass spectrometer with the exception of multicollection technology) for the analysis of transient peaks.

Improved Precision

To better understand how TOF technology can help improve precision in ICP-MS, it is important to know the major sources of instability. The most common source of noise in ICP-MS is flicker noise associated with the sample introduction process (peristaltic pump pulsations, nebulization mechanisms, plasma fluctuations, etc.) and shot noise derived from photons, electrons, and ions hitting the detector. Shot noise is based on counting statistics and is directly proportional to the square root of the signal. It therefore follows that as the signal intensity gets larger, the shot noise has less of an impact on the precision (% RSD) of the signal. This means that at high ion counts, the most dominant source of imprecision in ICP-MS is derived from flicker noise generated in the sample introduction area.

One of the most effective ways to reduce instability produced by flicker noise is to use a technique called internal standardization, where the analyte signal is compared and ratioed to the signal of an internal standard element (usually of a similar mass and/or ionization characteristics) that is spiked into the sample. Although a quadrupole-based system can do an adequate job of compensating for these signal fluctuations, it is ultimately limited by its inability to measure the internal standard at exactly the same time as the analyte isotope. So in order to compensate for sample introduction- and plasma-based noise and achieve high precision, the analyte and internal standard isotopes need to be sampled and measured simultaneously. For this reason, the design of a TOF mass analyzer is ideal for true simultaneous internal standardization required for high-precision work. It follows therefore that TOF is also well suited for high-precision isotope ratio analysis where its simultaneous nature of measurement is capable of achieving precision values close to the theoretical limits of counting statistics. In addition, unlike a scanning quadrupole-based system, it can measure ratios for as many isotopes or isotopic pairs as needed—all with excellent precision (8).

Rapid Data Acquisition

Like a scanning ICP-OES system, the speed of a quadrupole ICP mass spectrometer is limited by its scanning rate. To determine 10 elements in duplicate with good precision and detection limits, an integration time of 3 sec per mass is normally required. When overhead scanning and settling times are added for each mass and each replicate, this translates to approximately 2 min per sample. With a TOF system, the same analysis would take significantly less time because all the data are captured simultaneously. In fact, detection limit levels in a TOF instrument are typically achieved using a 10–30 sec integration time, which translates into a 5–10× improvement in data acquisition time over a quadrupole instrument. The added benefit of a TOF instrument is that speed of the analysis is not impacted by the number of analytes being determined. It would not matter if the method contained 10 or 70 elements—the measurement time would be virtually the same. However, there is one point that must be stressed. A large portion of the overall analysis time is taken up with flushing an old sample out and pumping a new sample into the sample introduction system. This can be as much as 2 min per sample for real-world matrices. So when this is taken into account, the difference between the sample throughput of a quadrupole and a TOF-ICP mass spectrometer is not so evident.

There is no question that TOF-ICP-MS, with its rapid, simultaneous mode of measurement, excels at multielement applications that generate fast transient signals such as laser ablation. It offers excellent precision, particularly for isotope ratioing techniques, and also has the potential for very fast data acquisition. However, this approach was only commercialized in 1998, so it is relatively immature compared to quadrupole ICP-MS technology. For that reason, it might need a little more time before it is ready for the severe demands of a routine, high-throughput laboratory.

FURTHER READING

1. Cameron AE, Eggers DF. Rev Sci Instrum 1948; 19(9):605.
2. Myers DP, Li G, Yang P, Hieftje GM. J Am Soc Mass Spectrom,1994; 5:1008–1016.
3. Myers DP. 12th Asilomar Conference on Mass Spectrometry, Pacific Grove, CA, Sept. 20–24, 1996.
4. Technical Note: 001-0877-00, GBC Scientific, February 1998.
5. Mahoney P, Li G, Hieftje GM. J Am Soc Mass Spectrom 1996; 11:401–406.
6. Technical Note: 001-0876-00, GBC Scientific, February 1998.
7. Sturgeon RE, Lam JWH, Saint A. J Anal At Spectrom 2000; 15:607–616.
8. Vanhaecke F, Moens L, Dams R, Allen L, Georgitis S. Anal Chem 1999; 71:3297.

10

Mass Analyzers: Collision/Reaction Cell Technology

The detection capability for some elements using traditional quadrupole mass analyzer technology is severely compromised because of the formation of polyatomic spectral interferences generated by a combination of argon, solvent, and/or sample-based ionic species. Although there are ways to minimize these interferences including correction equations, cool plasma technology, and matrix separation, they cannot be completely eliminated. However, a new approach called collision/reaction cell technology has recently been developed which virtually stops the formation of many of these harmful species before they enter the mass analyzer. This chapter takes a detailed look at this innovative new technique and the exciting potential it has to offer.

There are a small number of elements, which are recognized as having poor detection limits by ICP-MS. These are predominantly elements that suffer from major spectral interferences generated by ions derived from the plasma gas, matrix components, or the solvent/acid used to get the sample into solution. Examples of these interferences include:

→ $^{40}Ar^{16}O^+$ on the determination of $^{56}Fe^+$
→ $^{38}ArH^+$ on the determination of $^{39}K^+$
→ $^{40}Ar^+$ on the determination of $^{40}Ca^+$
→ $^{40}Ar^{40}Ar^+$ on the determination of $^{80}Se^+$
→ $^{40}Ar^{35}Cl^+$ on the determination of $^{75}As^+$
→ $^{40}Ar^{12}C^+$ on the determination of $^{52}Cr^+$
→ $^{35}Cl^{16}O^+$ on the determination of $^{51}V^+$

The cold/cool plasma approach, which uses a lower temperature to reduce the formation of the argon-based interferences, has been a very effective way to get around some of these problems (1). However, it can sometimes be difficult to optimize, is only suitable for a few of the interferences, is susceptible to more severe matrix effects, and can be time-consuming changing back and forth between normal and cool plasma conditions. These limitations and the desire to improve performance led to the development of collision/reaction cells in the late 1990s. Designed originally for organic MS to generate daughter species in order to confirm identification of the structure of the parent molecule (2), they found a use in ICP-MS to stop the formation of many argon-based spectral interferences.

BASIC PRINCIPLES OF COLLISION/REACTION CELLS

With this approach, ions enter the interface in the normal manner, where they are extracted into a collision/reaction cell under vacuum, which is positioned prior to the analyzer quadrupole. A collision/reaction gas such as hydrogen or helium is then bled into the cell, which consists of a multipole (a quadrupole, hexapole, or octapole), usually operated in the RF-only mode. The RF-only field does not separate the masses like a traditional quadrupole, but instead has the effect of focusing the ions, which then collide and react with molecules of the collision/reaction gas. By a number of different ion–molecule collision and reaction mechanisms, polyatomic interfering ions like $^{40}Ar^+$, $^{40}Ar^{16}O^+$, and $^{38}ArH^+$ will either be converted to harmless noninterfering species or the analyte will be converted to another ion which is not interfered with. This is exemplified by the reaction below, which shows the use of hydrogen gas to reduce the $^{38}ArH^+$ polyatomic interference in the determination of $^{39}K^+$. It can be seen that hydrogen gas converts $^{38}ArH^+$ to the harmless H_3^+ ion and atomic argon, but does not react with the potassium. The $^{39}K^+$ analyte ions, free of the interference, then emerge from the collision/reaction cell, where they are directed towards the quadrupole analyzer for normal mass separation.

$$\rightarrow\ ^{38}ArH^+ + H_2 = H_3^+ + Ar$$
$$\rightarrow\ ^{39}K^+ + H_2 = {}^{39}K^+ + H_2 \text{ (no reaction)}$$

The layout of a typical collision/reaction cell instrument is shown in Figure 10.1.

DIFFERENT COLLISION/REACTION APPROACHES

The above example is a very simplistic explanation of how a collision/reaction cell works. In practice, complex secondary reactions and collisions take place, which generate many undesirable interfering species. If these species were not eliminated or rejected, they could potentially lead to ad-

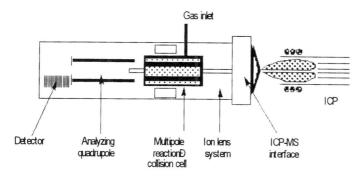

Gas inlet

ICP

Detector   Analyzing   Multipole   Ion lens   ICP-MS
           quadrupole   reaction/   system   interface
                    collision cell

FIGURE 10.1   Layout of a typical collision/reaction cell instrument.

ditional spectral interferences. There are basically two different approaches used to reject the products of these unwanted interactions.

→ Discrimination by kinetic energy
→ Discrimination by mass filtering

The major differences between the two approaches are in the types of multipoles used and their basic mechanism for rejection of the interferences. Let us take a closer look at how they differ.

Discrimination by Kinetic Energy

The first commercial collision cells for ICP-MS were based on hexapole technology (3), which was originally designed for the study of organic molecules using tandem mass spectrometry. The more collision-induced daughter species that were generated, the better the chance of identifying the structure of the parent molecule. However, this very desirable characteristic for LC or electrospray MS studies was a disadvantage in inorganic mass spectrometry, where secondary reaction-product ions are something to be avoided. There were ways to minimize this problem, but they were still limited by the type of collision gas that could be used. Unfortunately, highly reactive gases, such as ammonia and methane, which are more efficient at interference reduction, could not be used because of the limitations of a nonscanning hexapole (in RF-only mode) to adequately control the secondary reactions. The fundamental reason is that hexapoles do not provide adequate mass discrimination capabilities to suppress the unwanted secondary reactions, which necessitates the need for kinetic energy discrimination to distinguish the collision product ions from the analyte ions. Kinetic energy discrimination is typically achieved by setting the collision cell potential slightly more negative than the mass filter potential. This means that the collision-product ions generated in the cell, which have a lower kinetic energy as a result of the collision process, are rejected, while the analyte ions, which have a higher kinetic energy, are transmitted.

The inability of a hexapole-based collision cell to adequately control the secondary reactions therefore meant that low-reactivity gases like He, H_2, and Xe were the only option. The result was that ion–molecule collisional fragmentation (and not reactions) became the dominant mechanism of interference reduction. So although ion transmission characteristics of a hexapole were considered very good, detection limits were still relatively poor because the interference reduction process using hydrogen gas was much less efficient than using ammonia. For this reason, the performance of a kinetic energy-based collision cell, particularly for some of the more difficult elements, like Fe, K, and Ca, offered little improvement over the cool plasma approach. Table 10.1 shows some typical sensitivities (cps/ppm)

TABLE 10.1 Typical Sensitivities and Detection Limits Achievable with a Hexapole-Based Collision Cell ICP-MS

Element/mass	Sensitivity (cps/ppm)	Detection limit (ppt)
$^9Be^+$	6.9×10^7	7.7
$^{24}Mg^+$	1.3×10^8	28
$^{40}Ca^+$	2.8×10^8	70
$^{51}V^+$	1.7×10^8	0.9
$^{52}Cr^+$	2.4×10^8	0.7
$^{55}Mn^+$	3.4×10^8	1.7
$^{56}Fe^+$	3.0×10^8	17
$^{59}Co^+$	2.7×10^8	0.7
$^{60}Ni^+$	2.1×10^8	16
$^{63}Cu^+$	1.9×10^8	3
$^{68}Zn^+$	1.1×10^8	8
$^{88}Sr^+$	4.9×10^8	0.3
$^{107}Ag^+$	3.5×10^8	0.3
$^{114}Cd^+$	2.4×10^8	0.4
$^{128}Te^+$	1.3×10^8	9
$^{138}Ba^+$	5.9×10^8	0.2
$^{205}Tl^+$	4.0×10^8	0.2
$^{208}Pb^+$	3.7×10^8	0.7
$^{209}Bi^+$	3.4×10^8	0.5
$^{238}U^+$	2.3×10^8	0.1

Source: Ref. 4.

and detection limits (ppt) achievable with a hexapole-based collision cell ICP-MS (4).

Recent modifications to the hexapole design have significantly improved its collision/reaction characteristics. In addition to offering good transmission characteristics and kinetic energy discrimination, they now appear to offer basic mass-dependent discrimination capabilities. This means that the kinetic energy discrimination barrier can be adjusted with analytical mass, which offers the capability of using small amounts of highly reactive gases. This is exemplified in Figure 10.2, which shows the reduction of both $^{40}Ar^{12}C^+$ and $^{37}Cl^{16}O^+$ using helium with a small amount of ammonia in the isotopic ratio determination of $^{52}Cr^+/^{53}Cr^+$ (^{52}Cr is 83.789% and ^{53}Cr is 9.401% abundant). It can be seen that the $^{52}Cr^+/^{53}Cr^+$ ratio is virtually the same in the chloride and carbon matrices as it is with no matrix present when the optimum flow of collision/reaction gas is used (5).

Another approach to using a hexapole is to utilize an octapole in the collision cell. The benefit of using a higher-order design is that its trans-

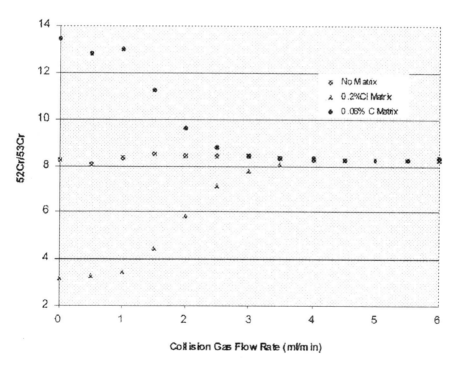

FIGURE 10.2 The use of helium/ammonia mixture with a hexapole-based collision cell for the successful determination of $^{52}Cr^+/^{53}Cr^+$ isotopic ratios. (From Ref. 5.)

mission characteristics, particularly at the low mass end, are slightly higher than lower-order multipoles. Similar in design to the hexapole, collisional fragmentation and energy discrimination are the predominant mechanisms for interference reduction, which means that lower reactivity gases like hydrogen and helium are preferred. By careful design of the interface and the entrance to the cell, the collision/reaction capabilities can be improved by reducing the number of sample/solvent/plasma-based ions entering the cell. This enables the collision gas to be more effective at reducing the interferences. An example of this is the use of H_2 as the cell gas to reduce the argon dimer ($^{40}Ar_2^+$) interference in the determination of the major isotope of selenium at mass 80 ($^{80}Se^+$). This is exemplified in Figure 10.3, which shows a dramatic reduction in the $^{40}Ar_2^+$ background at mass 80, using an ICP-MS fitted with an octapole reaction cell. It can be seen that by using the optimum flow of H_2, the spectral background has been reduced by about 6 orders of magnitude from 10 million to 10 cps, producing a BEC of approximately 1 ppt for $^{80}Se^+$ (6).

Discrimination by Mass Filtering

Another way of rejecting the products of the secondary reactions/collisions is to discriminate them by mass. Unfortunately, higher-order multipoles cannot be used for efficient mass discrimination because the stability boundaries are diffuse and the sequential secondary reactions cannot be easily intercepted. The way around this is to utilize a quadrupole (instead of a hexapole or octapole) inside the reaction/collision cell and use it as a selective band-pass (mass) filter. The benefit of this approach is that highly re-

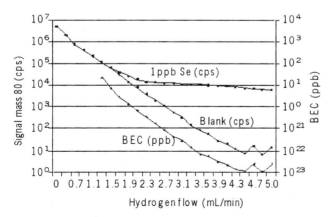

FIGURE 10.3 Background reduction of the argon dimer ($^{40}Ar_2^+$) with hydrogen gas using an octapole reaction cell. (From Ref. 6.)

active gases can be used, which tend to be more efficient at interference reduction. One such development that uses this approach is called dynamic reaction cell technology (DRC) (7,8). Similar in appearance to the hexapole and octapole collision/reaction cells, the dynamic reaction cell is a pressurized multipole positioned prior to the analyzer quadrupole. However, this is where the similarity ends. In DRC technology, a quadrupole is used instead of a hexapole or octapole. A highly reactive gas, such as ammonia or methane, is bled into the cell, which is a catalyst for ion molecule chemistry to take place. By a number of different reaction mechanisms, the gaseous molecules react with the interfering ions to convert them either into a innocuous species different to the analyte mass or a harmless neutral species. The analyte mass then emerges from the dynamic reaction cell free of its interference and steered into the analyzer quadrupole for conventional mass separation. The advantage of using a quadrupole in the reaction cell is that the stability regions are much better defined than a hexapole or an octapole, so it is relatively straightforward to operate the quadrupole inside the reaction cell as a mass or band-pass filter, and not just an ion-focusing guide. Therefore by careful optimization of the quadrupole electrical fields, unwanted reactions between the gas and the sample matrix or solvent, which could potentially lead to new interferences, are prevented. This means that every time an analyte and interfering ions enter the dynamic reaction cell, the band pass of the quadrupole can be optimized for that specific problem and then changed on the fly for the next one. This is seen schematically in Figure 10.4, which shows an analyte ion ^{56}Fe and an isobaric interference $^{40}Ar^{16}O^+$ entering the dynamic reaction cell. As can be seen, the reaction

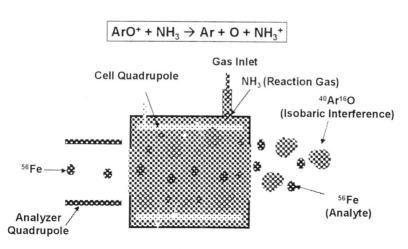

FIGURE 10.4 Elimination of the $^{40}Ar^{16}O^+$ interference with a dynamic reaction cell.

gas NH_3 reacts with the ArO^+ to form atomic oxygen and argon together with a positive NH_3 ion. The quadrupole's electrical field is then set to allow the transmission of the analyte ion ^{56}Fe to the analyzer quadrupole, free of the problematic isobaric interference, $^{40}Ar^{16}O^+$. In addition, the NH_3^+ is prevented from reacting further to produce a new interfering ion. The benefit of this approach is that highly reactive gases can be used which increases the number of ion–molecule reactions taking place and therefore more efficient removal of the interfering species. Of course, this also potentially generates more side reactions between the gas and the sample matrix and solvent. However, by dynamically scanning the band pass of the quadrupole in the reaction cell, these reaction by-products are rejected before they can react to form new interfering ions.

The benefit of the DRC approach is that by careful selection of the reaction gas, an advantage can be taken of the different rates of reaction of the analyte and the interfering species. This can be exemplified by the elimination of $^{40}Ar^+$ by NH_3 gas in the determination of $^{40}Ca^+$. The reaction between NH_3 gas and the $^{40}Ar^+$ interference, which is predominantly charge exchange, occurs because the ionization potential of NH_3 (10.2 eV) is low compared to that of Ar (15.8 eV). This makes the reaction exothermic and fast. However, the ionization potential of Ca (6.1 eV) is significantly less than that of NH_3, so the reaction, which is endothermic, is not allowed to proceed (8). This can be seen in greater detail in Figure 10.5.

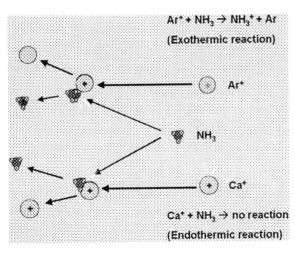

FIGURE 10.5 The reaction between NH_3 and Ar^+ is exothermic and fast, while there is no reaction between NH_3 and Ca^+ in the dynamic reaction cell.

This highly efficient reaction mechanism translates into a dramatic reduction of the spectral background at mass 40, which is shown graphically in Figure 10.6. It can be seen that at the optimum NH_3 flow, a reduction in the $^{40}Ar^+$ background signal of about 8 orders of magnitude is achieved, resulting in a detection limit of < 0.5 ppt for $^{40}Ca^+$.

It should be pointed out that although a multipole collision cell using hydrogen gas (with energy discrimination) can be as efficient to reduce the $^{40}Ar^+$ background, it requires significantly more collisions than a reaction cell that uses a highly reactive gas (9). This alone is not such a problem, but it must be remembered that the $^{40}Ca^+$ will also lose kinetic energy because it experiences the same number of collisions as the $^{40}Ar^+$ ion. This means that the transmission of $^{40}Ca^+$ ions to the mass analyzer will also be reduced because of the potential energy barrier downstream of the cell, resulting in a compromised detection limit compared to a reaction cell system.

Table 10.2 shows some typical detection limits in parts per trillion (ppt) of an ICP-MS system fitted with a dynamic reaction cell (10). The elements with an asterisk (*) were determined using ammonia as the reaction gas, while the other elements were determined in the standard mode (no reaction gas).

There is no question that collision/reaction cells have given a new lease of life to quadrupole mass analyzers used in ICP-MS. They have enhanced its performance and flexibility and most definitely opened up the technique to more demanding applications, which were previously beyond its capa-

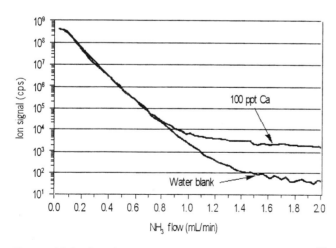

FIGURE 10.6 A reduction of 8 orders of magnitude in the $^{40}Ar^+$ background signal is achievable with the dynamic reaction cell—resulting in <0.5 ppt detection limit for $^{40}Ca^+$.

TABLE 10.2 Typical Detection Limits in Parts per Trillion of an ICP-MS Fitted with a Dynamic Reaction Cell

Element	DL (ppt)	Element	DL (ppt)
As	0.48	Fe*	0.12
B	1.93	Ni	0.10
Na	0.14	Co*	0.04
Mg	0.08	Cu*	0.05
Al*	0.05	Zn*	0.45
K*	0.27	Sn	0.12
Ca*	0.10	Sb	0.08
Ti*	0.92	Ba	0.06
V*	0.12	Pb	0.07
Cr*	0.12	Ni	0.10
Mn*	0.17		

* Determined using NH_3 as the reaction gas.
Source: Ref. 10.

bilities. However, it must be emphasized that although there are differences between commercially available instruments, they all perform extremely well. The intent of this chapter is to present the overall benefits of the technology and give you an overview of the different approaches available. If it has created an interest, I strongly suggest that a performance evaluation is made based on your own application problems.

FURTHER READING

1. Sakata K, Kawabata K. Spectrochim Acta 1994; 49B:1027.
2. Thomson BA, Douglas DJ, Corr JJ, Hager JW, Joliffe CA. Anal Chem 1995; 67:1696–1704.
3. Turner P, Merren T, Speakman J, Haines C. Plasma Source Mass Spectrometry: Developments and Applications. Cambridge, England: Royal Society of Chemistry, 1996:28–34.
4. Feldmann I, Jakubowski N, Thomas C, Stuewer D. Fresenius' J Anal Chem 1999; 365:422–428.
5. Collision Cell Technology with Energy Discrimination, Thermo Elemental Application Note, September 2001.
6. McCurdy E, Potter D. Multielement Analysis of Unknown Sample Matrices with a Reaction Cell System, Agilent Technologies ICP-MS Journal—Issue 10, October 2001.
7. Covered by US Patent No. 6140638.
8. Tanner SD, Baranov VI. At Spectr 1999; 20(2):45–52.
9. Tanner SD, Baranov VI, Voellkopf U. J Anal At Spectrom 2000; 15:1261–1269.
10. Kawabata K, Kishi Y, Thomas R. Spectroscopy 2003; 18(1):2–9.

11

Detectors

This chapter takes a look at the detection system—an important area of the mass spectrometer that quantifies the number of ions emerging from the mass analyzer. The detector converts the ions into electrical pulses, which are then counted using its integrated measurement circuitry. The magnitude of the electrical pulses corresponds to the number of analyte ions present in the sample, which is then used for trace element quantitation by comparing the ion signal with known calibration or reference standards.

Since ICP-MS was first introduced in the early 1980s, a number of different ion detection designs have been utilized, the most popular being electron multipliers for low ion count rates and faraday collectors for high count rates. Today, the majority of ICP-MS systems that are used for ultratrace analysis use detectors that are based on the active film or discrete dynode electron multiplier. They are very sophisticated pieces of equipment and are very efficient at converting ion currents emerging from the mass analyzer into electrical signals. The location of the detector in relation to the mass analyzer is shown in Figure 11.1.

Before we go on to describe discrete dynode detectors in greater detail, it is worth looking at two of the earlier designs—the channel electron multiplier (Channeltron®) (1) and the Faraday cup—in order to get a basic understanding of how the ICP-MS ion detection process works.

CHANNEL ELECTRON MULTIPLIER

The operating principles of the channel electron multiplier are similar to a photomultiplier tube used in ICP-OES. However, instead of using individual dynodes to convert photons to electrons, the channeltron is an open glass cone (coated with a semiconductor-type material) to generate electrons from ions that impinge on its surface. For the detection of positive ions, the front

91

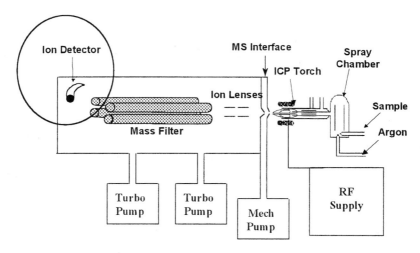

FIGURE 11.1   The location of the detector in relation to the mass analyzer.

of the cone is biased at a negative potential, while the far end near the collector is kept at ground. When the ion emerges from the quadrupole mass analyzer, it is attracted to the high negative potential of the cone. When the ion hits this surface, one or more secondary electrons are formed. The potential gradient inside the tube varies based on position, so the secondary electrons move further down the tube. As these electrons strike new areas of the coating, more secondary electrons are emitted. This process is repeated many times. The result is a discrete pulse, which contains millions of electrons generated from an ion that first hits the cone of the detector (1). This process is shown simplistically in Figure 11.2.

This pulse is then sensed and detected by a very fast preamplifier. The output pulse from the preamplifier then goes to a digital discriminator and counting circuitry, which only counts pulses above a certain threshold value. This threshold level needs to be high enough to discriminate against pulses caused by spurious emission inside the tube, from any stray photons from the plasma itself, or photons generated from fast-moving ions striking the quadrupole rods.

It is worth pointing out that the rate of ions hitting the detector is sometimes too fast for the measurement circuitry to handle them in an efficient manner. This is caused by ions arriving at the detector during the output pulse of the preceding ion and not being detected by the counting system. This "dead time," as it is known, is a fundamental limitation of the multiplier detector and is typically 30–50 nsec, depending on the detection

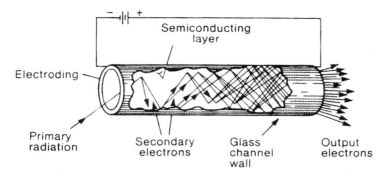

FIGURE 11.2   Channel electron multiplier. (From Ref. 1.)

system. Compensation in the measurement circuitry has to be made for this "dead time" in order to count the maximum number of ions hitting the detector.

FARADAY CUP

For some applications, where ultratrace detection limits are not required, the ion beam from the mass analyzer is directed into a simple metal electrode or Faraday cup. With this approach, there is no control over the applied voltage (gain), so they can only be used for high ion currents. Their lower working range is in the order of 10^4 cps, which means that if they are to be used as the only detector, the sensitivity of the ICP mass spectrometer will be severely compromised. For this reason, they are normally used in conjunction with a channeltron or discrete dynode detector to extend the dynamic range of the instrument. An additional problem with the Faraday cup is that because of the time constant used in the DC amplification process to measure the ion current, they are limited to relatively low scan rates. This limitation makes them unsuitable for the fast scan rates required for traditional pulse counting used in ICP-MS and also limits their ability to handle fast transient peaks.

The Faraday cup never became popular with quadrupole ICP-MS systems because it was not suitable for very low ion count rates. An attempt was made in the early 1990s to develop an ICP-MS system using a Faraday cup detector for the environmental market, but its sensitivity was compromised, and, as a result, was considered more suitable for applications requiring ICP-OES trace level detection capability. However, Faraday cup technology is still utilized in some magnetic sector instruments, particularly where high ion signals are encountered in the determination of high-precision isotope ratios, using a multicollector detection system.

DISCRETE DYNODE ELECTRON MULTIPLIER

These detectors, which are often called active film multipliers, work in a similar way to the channeltron, but utilize discrete dynodes to carry out the electron multiplication (2). Figure 11.3 illustrates the principles of operation of this device. The detector is positioned off-axis to minimize the background noise from stray radiation and neutral species coming from the ion source. When an ion emerges from the quadrupole, it sweeps through a curved path before it strikes the first dynode. On striking the first dynode, it liberates secondary electrons. The electro-optic design of the dynode produces acceleration of these secondary electrons to the next dynode where they generate more electrons. This process is repeated at each dynode, generating a pulse of electrons that are finally captured by the multiplier collector or anode. Because of the materials used in the discrete dynode detector and the difference in the way electrons are generated, it is typically 50–100% more sensitive than channeltron technology.

Although most discrete dynode detectors are very similar in the way they work, there are subtle differences in the way the measurement circuitry handles low and high ion count rates. When ICP-MS was first commercial-

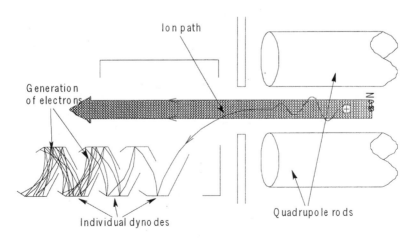

Figure 11.3   Schematic of a discrete dynode electron multiplier. (From Ref. 2.)

ized, it could only handle up to 5 orders of dynamic range. However, when attempts were made to extend the dynamic range, certain problems were encountered. Before we discuss how modern detectors deal with this issue, let us first take a look at how it was addressed in earlier instrumentation.

EXTENDING THE DYNAMIC RANGE

Traditionally, ICP-MS using the pulse-counting measurement is capable of about 5 orders of linear dynamic range. This means that ICP-MS calibration curves, generally speaking, are linear from detection limit up to a few hundred parts per billion. However, there are a number of ways of extending the dynamic range of ICP-MS another 3–4 orders of magnitude and working from sub-parts per thousand levels up to a hundred parts per million. Here is a brief overview of some of different approaches that have been used.

Filtering the Ion Beam

One of the very first approaches to extend the dynamic range in ICP-MS was to filter the ion beam. This was achieved by putting a nonoptimum voltage on one of the ion lens components or the quadrupole itself to limit the number of ions reaching the detector. This voltage offset, which was set on an individual mass basis, acted as an energy filter to electronically screen the ion beam and reduce the subsequent ion signal to within a range covered by pulse-counting ion detection. The main disadvantage with this approach was that the operator had to have prior knowledge of the sample to know what voltage to apply to the high-concentration masses.

Using Two Detectors

Another technique that was used on some of the early ICP-MS instrumentation was to utilize two detectors, such as a channel electron multiplier and a Faraday cup, to extend the dynamic range. With this technique, two scans would be made. In the first scan, it would measure the high-concentration masses using the Faraday cup, then in the second scan, it would skip over the high-concentration masses and carry out pulse counting of the low concentration masses with a channel electron multiplier. This worked reasonably well but struggled with applications that required rapid switching between the two detectors because the ion beam had to be physically deflected in order to select the optimum detector. Not only did this degrade the measurement duty cycle, but detector switching and stabilization times of several seconds also precluded fast transient signal detection.

Using Two Scans with One Detector

The more modern approach is to use just one detector to extend the dynamic range. This has typically been done by using the detector both in pulse and analog mode, so high and low concentrations can be determined in the same sample. There are basically three approaches using this type of detection system—two of them involve carrying out two scans of the sample, while the third uses only one scan.

The first approach uses an electron multiplier operated in both digital and analog modes (3). Digital counting provides the highest sensitivity, while operation in the analog mode (achieved by reducing the high voltage applied to the detector) is used to reduce the sensitivity of the detector, thus extending the concentration range for which ion signals can be measured. The system is implemented by scanning the spectrometer twice for each sample. The first scan, in which the detector is operated in the analog mode, provides signals for elements present at high concentrations. A second scan, in which the detector voltage is switched to digital, pulse-counting mode, provides high-sensitivity detection for elements present at low levels. A major advantage of this technology is that the user does not need to know in advance whether to use analog or digital detection because the system automatically scans all elements in both modes. However, its major disadvantage is that two independent mass scans are required in order to gather data across an extended signal range. This not only results in degraded measurement efficiency and slower analyses, but it also means that the system cannot be used for fast transient signal analysis because mode switching is generally too slow.

An alternative way of extending the dynamic range is similar to the first approach, except that the first scan is used as an investigative tool to examine the sample spectrum before analysis (4). This first prescan establishes the mass positions at which the analog and pulse modes will be used for subsequently collecting the spectral signal. The second analytical scan is then used for data collection, switching the detector back and forth rapidly between pulse and analog mode at each analytical mass.

Although these three approaches work very well, their main disadvantage is that two separate scans are required to measure high and low levels. With conventional nebulization, this is not such a major problem except that it can impact sample throughput. However, it does become a concern when it comes to working with transient peaks found in electrothermal vaporization (ETV), flow injection (FIAS), or laser sampling (LS) ICP-MS. Because these transient peaks often only last a few seconds, all the available time must be spent measuring the masses of interest to get the best detection limits. When two scans have to be made, time is wasted collecting data, which is not contributing to the analytical signal.

Using One Scan with One Detector

This limitation of having to scan twice led to the development of an alternative design using a dual-stage discrete dynode detector (5). This technology utilizes measurement circuitry that allows both high and low concentrations to be determined in one scan. This is achieved by measuring the ion signal as an analog signal at the midpoint dynode. When more than a threshold number of ions are detected, the signal is processed through the analog circuitry. When fewer than the threshold number of ions is detected, the signal cascades through the rest of the dynodes and is measured as a pulse signal in the conventional way. This process, which is shown in Figure 11.4, is completely automatic and means that both the analog and the pulse signals are collected simultaneously in one scan (6).

The pulse-counting mode is typically linear from 0 to about 10^6 cps, while the analog circuitry is suitable from 10^4 to 10^9 cps. To normalize both ranges, a cross-calibration is carried out to cover concentration levels, which produce a pulse and an analog signal. This is possible because the analog and pulse outputs can be defined in identical terms of incoming pulse counts per second, based on knowing the voltage at the first analog stage, the output current, and a conversion factor defined by the detection circuitry electron-

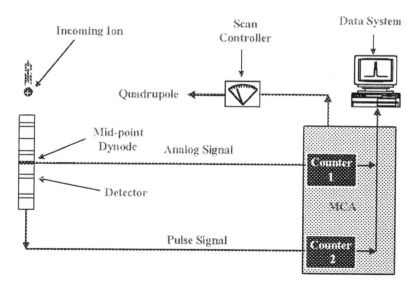

FIGURE 11.4 Dual-stage discrete dynode detector measurement circuitry. (From Ref. 5.)

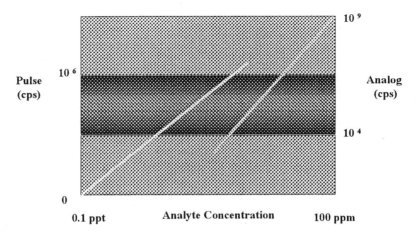

FIGURE 11.5 The pulse-counting mode covers up to 10^6 cps, while the analog circuitry is suitable from 10^4 to 10^9 cps, with a dual-mode discrete dynode detector. (From Ref. 5.)

ics. By carrying out a cross-calibration across the mass range, a dual mode detector of this type is capable of achieving approximately 8–9 orders of dynamic range in one simultaneous scan. This can be seen in Figures 11.5 and 11.6. Figure 11.5 shows that the pulse-counting calibration curve (left-hand plot) is linear up to 10^6 cps, while the analog calibration curve (right-hand plot) is linear from 10^4 to 10^9 cps. Figure 11.6 shows that after cross-calibration, the two curves are normalized, which means that the detector is suitable for concentration levels between 0.1 ppt and 100 ppm— typically 8–9 orders of magnitude for most elements (5).

There are subtle variations of this type of detection system, but its major benefit is that it requires only one scan to determine both high and low concentrations. It therefore not only offers the potential to improve sample throughput, but also means that the maximum data can be collected on a transient signal that only lasts a few seconds. This will be described in greater detail in Chapter 12, where we will discuss different measurement protocols and peak integration routines.

Extending the Dynamic Range Using Pulse-Only Mode

The most recent development in extending the dynamic range is to use the pulse-only signal. This is achieved by monitoring the ion flux at one of the first few dynodes of the detector (before extensive electron multiplication

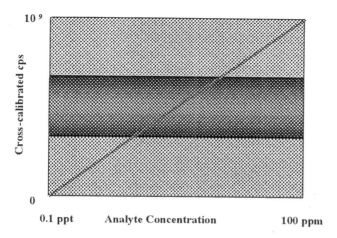

FIGURE 11.6 Using cross-calibration of the pulse and analog modes, quantitation from sub-parts per trillion to high parts per million levels is possible. (From Ref. 5.)

has taken place) and then attenuating the signal up to 10,000:1 by applying a control voltage. Electron pulses passed by the attenuation section are then amplified to yield pulse heights that are typical in normal pulse-counting applications (6).

There are basically three ways of implementing this technology based on the types of samples being analyzed. It can be run in conventional pulse-only mode for normal low-level work. It can also be run using an operator-selected attenuation factor if the levels of the higher concentration elements being determined are consistent and well understood. On the other hand, if the samples are completely unknown and have not been well characterized beforehand, a dynamic attenuation mode of operation is available. In this mode, an additional premeasurement time is built into the quadrupole settling time in order to determine the optimum detector attenuation for the selected dwell times used. It is unclear how much overhead this adds to the settling time, but it is probably less than 1 msec per scan.

This novel, pulse-only approach to extending the dynamic range looks to be a very exciting development, which does not suffer from the limitations of measuring both pulse and analog signals individually. However, it does require a preanalysis attenuation calibration to be carried out on a fairly frequent basis in order to determine the extent of signal attenuation required. The frequency of calibration is unknown at this time because there are very few instruments in the field carrying out real-world analysis. But based on current information supplied by the instrument manufacturer, it is expected to be in the order of 3–4 weeks.

FURTHER READING

1. Channeltron® Electron Multiplier Handbook for Mass Spectrometry Applications, Galileo Electro-Optic Corp., 1991. Channeltron is a registered trademark of Galileo Corp.
2. Hunter K. At Spectr 1994; 15(1):17–20.
3. Hutton RC, Eaton AN, Gosland RM. Appl Spectrosc 1990; 44(2):238–242.
4. Kishi Y. Agilent Technol Appl J, August 1997.
5. Denoyer ER, Thomas RJ, Cousins L. Spectroscopy 1997; 12(2):56–61.
6. Covered by US Patent Number 5,463,219.
7. Gray J, Stresau R, Hunter K. Ion Counting Beyond 10 GHz, Poster Presentation Number 890-6P. Orlando, FL: Pittsburgh Conference and Exposition, 2003.

12

Peak Measurement Protocol

With its multielement capability, superb detection limits, wide dynamic range, and high sample throughput, ICP-MS is proving to be a compelling technique for more and more diverse application areas. However, no two application areas have the same analytical requirements. For example, environmental and clinical contract labs, although wanting reasonably low detection limits, are not really pushing the technique to its extreme detection capability. Their main requirement is usually high sample throughput, because the number of samples these labs can analyze in a day directly impacts their revenue. On the other hand, a semiconductor fabrication plant or a supplier of high-purity chemicals to the electronics industry is interested in the lowest detection limits the technique can offer because of the contamination problems associated with manufacturing high-performance electronic devices.

To meet such diverse application needs, modern ICP-MS instrumentation has to be very flexible if it is to keep up with the increasing demands of its users. Nowhere is this more important than in the area of peak integration and measurement protocol. The way the analytical signal is managed in ICP-MS has a direct impact on its multielement characteristics, isotopic capability, detection limits, dynamic range, and sample throughput; the five major strengths that attracted the trace element community to the technique almost 20 years ago. To understand signal management in greater detail and its implications on data quality, we will discuss how measurement protocol is optimized based on the application's analytical requirements, and its impact on both continuous signals generated by traditional nebulization devices and transient signals produced by alternative sample introduction techniques such as flow injection and laser ablation.

MEASUREMENT VARIABLES

There are many variables that affect the quality of the analytical signal in ICP-MS. The analytical requirements of the application will often dictate this, but there is no question that instrumental detection and measurement parameters can have a significant impact on the quality of data in ICP-MS. Some of the variables that can potentially impact the quality of your data, particularly when carrying out multielement analysis, include:

- Whether it is a continuous or transient signal.
- The temporal length of the sampling event.
- Volume of sample available.
- Number of samples being analyzed.
- Number of replicates per sample.
- Number of elements being determined.
- Detection limits required.
- Precision/accuracy expected.
- Dynamic range needed.
- Integration time used.
- Peak quantitation routines.

Before we go on to discuss these in greater detail and how these parameters affect the data, it is important to remind ourselves how a scanning device such as a quadrupole mass analyzer works. Although we will focus on quadrupole technology, the fundamental principles of measurement protocol will be very similar for all types of mass spectrometers that use a scanning approach for multielement peak quantitation.

MEASUREMENT PROTOCOL

The principles of scanning with a quadrupole mass analyzer are shown in Figure 12.1. In this simplified example, the analyte ion (black) and four other ions have arrived at the entrance to the four rods of the quadrupole. When a particular RF/DC voltage is applied to each pair of rods, the positive or negative bias on the rods will electrostatically steer the analyte ion of interest down the middle of the four rods to the end, where it will emerge and be converted to an electrical pulse by the detector. The other ions of different mass to charge will pass through the spaces between the rods and be ejected from the quadrupole. This scanning process is then repeated for another analyte at a completely different mass-to-charge ratio until all the analytes in a multielement analysis have been measured.

The process for the detection of one particular mass in a multielement run is represented in Figure 12.2. It shows a $^{59}Co^+$ ion emerging from the

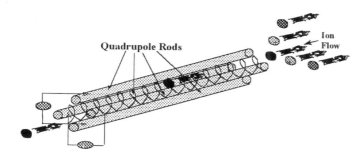

FIGURE 12.1 Principles of mass selection with a quadrupole mass filter.

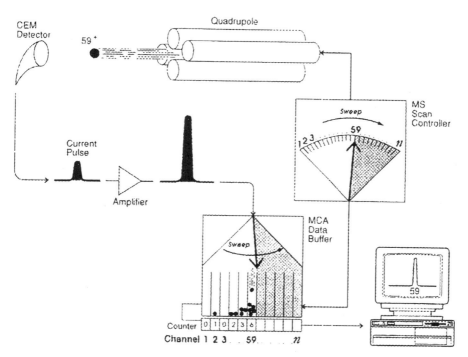

FIGURE 12.2 Detection and measurement protocol using a quadrupole mass analyzer. (From Ref. 1.)

quadrupole and being converted to an electrical pulse by the detector. As the optimum RF/DC ratio is applied for $^{59}Co^+$ and repeatedly scanned, the ions as electrical pulses are stored and counted by a multichannel analyzer. This multichannel data acquisition system typically has 20 channels per mass and as the electrical pulses are counted in each channel, a profile of the mass is built-up over the 20 channels, corresponding to the spectral peak of $^{59}Co^+$. In a multielement run, repeated scans are made over the entire suite of analyte masses, as opposed to just one mass represented in this example. The principles of multielement peak acquisition are shown in Figure 12.3. In this example (showing two masses), signal pulses are continually collected as the quadrupole is swept across the mass spectrum, shown by sweeps 1–3. After a fixed number of sweeps (determined by user), the total number of signal pulses in each channel is counted, resulting in the final spectral peak (1).

When it comes to quantifying an isotopic signal in ICP-MS, there are basically two approaches to consider. There is the multichannel ramp scanning approach, which uses a continuous smooth ramp of $1-n$ channels (where n is typically 20) per mass across the peak profile. This is shown in Figure 12.4.

And there is the peak hopping approach where the quadrupole power supply is driven to a discrete position on the peak (normally the maximum point), allowed to settle, and a measurement taken for a fixed amount of time. This is represented in Figure 12.5.

The multipoint scanning approach is best for accumulating spectral and peak shape information when doing mass scans. It is normally used for doing mass calibration and resolution checks and as a classical qualitative method development tool to find out what elements are present in the sample and to assess their spectral implications on the masses of interest. Full peak profiling is not normally used for doing rapid quantitative analysis, because valuable analytical time is wasted taking data on the wings and valleys of the peak, where the signal to noise is poorest.

When the best possible detection limits are required, the peak-hopping approach is best. It is important to understand that to get the full benefit of peak hopping, the best detection limits are achieved when single-point peak hopping at the peak maximum is chosen. However, to carry out single-point peak hopping it is essential that the mass stability is good enough to reproducibly go to the same mass point every time. If good mass stability can be guaranteed (usually by thermostating the quadrupole power supply), measuring the signal at the peak maximum will always give the best detection limits for a given integration time. It is well documented that there is no benefit to spread the chosen integration time over more than one measurement point per mass. If time is a major consideration in the analysis, then

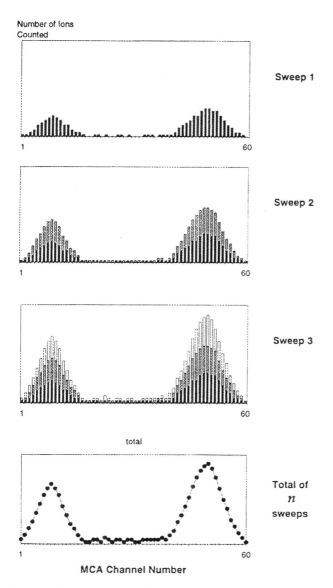

FIGURE 12.3 A profile of the peak is built up by continually sweeping the quadrupole across the mass spectrum. (From Ref. 1.)

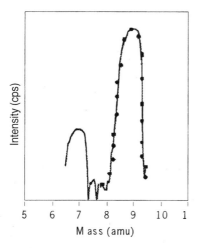

FIGURE 12.4   Multichannel ramp scanning approach using 20 channels per amu.

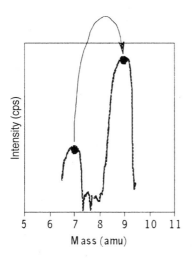

FIGURE 12.5   Peak hopping approach.

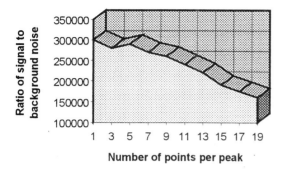

FIGURE 12.6 Signal to background noise (10 ppb rhodium) degrades when more than one point, spread over the same integration time, is used for peak quantitation.

using multiple points is wasting valuable time on the wings and valleys of the peak, which contribute less to the analytical signal and more to the background noise. This is shown in Figure 12.6, which shows the degradation in signal to background noise of 10 ppb Rh with an increase in the number of points per peak, spread over the same total integration time. Detection limit improvement for a selected group of elements using 1 point/peak compared to 20 points/peak is shown in Figure 12.7.

OPTIMIZATION OF MEASUREMENT PROTOCOL

Now that the fundamentals of the quadrupole measuring electronics have been described, let us now go into more detail on the impact of optimizing

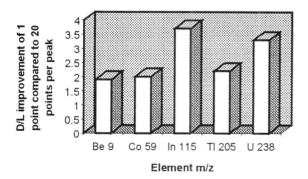

FIGURE 12.7 Detection limit improvement using 1 point/peak compared to 20 points/peak over the mass range. (From Ref. 2.)

the measurement protocol based on the requirement of the application. When multielement analysis is being carried out by ICP-MS, there are a number of decisions that need to be made. First, we need to know if we are dealing with a continuous signal from a nebulizer or a transient signal from an alternative sampling accessory. And if it is a transient event, how long will the signal last? Another question that needs to be addressed is how many elements are going to be determined? With a continuous signal, this is not such a major problem but could be an issue if we are dealing with a transient signal that lasts a few seconds. We also need to be aware of the level of detection capability required. This is a major consideration with a single-shot laser pulse that lasts 5–10 sec, but also with a continuous signal produced by a concentric nebulizer, we might have to accept a compromise of detection limit based on the speed of analysis requirements or amount of sample available. What analytical precision is expected? If isotope ratio/dilution work is being done, how many ions do we have to count to guarantee good precision? Does increasing the integration time of the measurement help the precision? Finally, is there a time constraint on the analysis? A high throughput laboratory might not be able to afford to use the optimum sampling time to get the ultimate in detection limit. In other words, what compromises need to be made between detection limit, precision, and sample throughput? It is clear that before the measurement protocol can be optimized, the major analytical requirements of the application need to be defined. Let us take a look at this in greater detail.

MULTIELEMENT DATA QUALITY OBJECTIVES

Because multielement detection capability is probably the major reason why most laboratories invest in ICP-MS, it is important to understand the impact of measurement criteria on detection limits. We know that in a multielement analysis, the quadrupole's RF/DC ratio is "driven" or scanned to mass regions, which represent the elements of interest. The electronics are allowed to settle and then "sit" or dwell on the peak and take measurements for a fixed period of time This is usually performed a number of times until the total integration time is fulfilled. For example, if a dwell time of 50 msec is selected for all masses and the total integration time is 1 sec, then the quadrupole will carry out 20 complete sweeps per mass, per replicate. It will then repeat the same routine for as many replicates that have been built into the method. This is seen very simplistically in Figure 12.8, which shows the scanning protocol of a multielement scan of three different masses.

In this example, the quadrupole is scanned to mass A. The electronics are allowed to settle (settling time), left to dwell for a fixed period of time at one or multiple points on the peak (dwell time), and intensity measurements

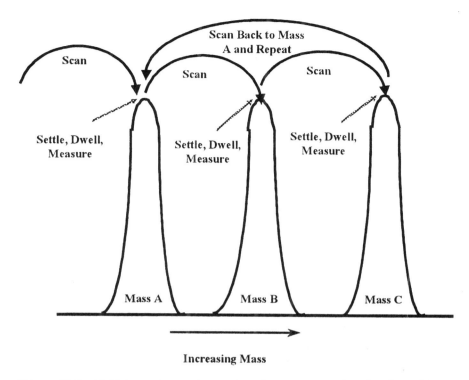

FIGURE 12.8 Multielement scanning and peak measurement protocol used in a quadrupole.

taken (based on the dwell time). The quadrupole is then scanned to masses B and C and the measurement protocol repeated. The complete multielement measurement cycle (sweep) is repeated as many times as needed to make up the total integration per peak. It should be emphasized that this is a generalization of the measurement routine—management of peak integration by the software will vary slightly based on different instrumentation.

It is clear from this that during a multielement analysis there is a significant amount of time spent scanning and settling the quadrupole, which does not contribute to the quality of the analytical signal. Therefore if the measurement routine is not optimized carefully, it can have a negative impact on data quality. The dwell time can usually be selected on an individual mass basis, but the scanning and settling times are normally fixed because they are a function of the quadrupole and detector electronics. For this reason, it is essential that the dwell time, which ultimately affects detection limit and precision, must dominate the total measurement time, compared to the scanning, and settling times. It follows therefore that the measurement duty cycle (% of actual measuring time compared to total integration time) is maximized when the quadrupole and detector electronics settling times are kept to an absolute minimum. This can be seen in Figure 12.9, which shows a plot of % measurement duty cycle against dwell time for four different quadrupole settling times—0.2, 1.0, 3.0, and 5.0 msec for one replicate of a multielement scan of five masses, using one point per peak. In this example, the total integration time for each mass was 1 sec, with the number of sweeps varying depending on the dwell time used. For this exercise, the % duty cycle is defined by the following equation:

$$\frac{\text{Dwell Time} \times \# \text{Sweeps} \times \#\text{Elements} \times \# \text{Replicates}}{\{(\text{Dwell time} \times \# \text{Sweeps} \times \# \text{Elements} \times \# \text{Replicates}) + (\text{Scanning/Settling Time} \times \# \text{Sweeps} \times \# \text{Elements} \times \# \text{Replicates})\}} \times 100$$

So in order to achieve the highest duty cycle, the nonanalytical time must be kept to an absolute minimum. This leads to more time being spent counting ions and less time scanning and settling, which do not contribute to the quality of the analytical signal. This becomes of critical importance when a rapid transient peak is being quantified, because the available measuring time is that much shorter (3). Generally speaking, peak quantitation using multiple points per peak and long settling times should be avoided in ICP-MS because it ultimately degrades the quality of the data for a given integration time.

It can also be seen in Figure 12.9 that shorter dwell times translate into a lower duty cycle. For this reason, for normal quantitative analysis work, it

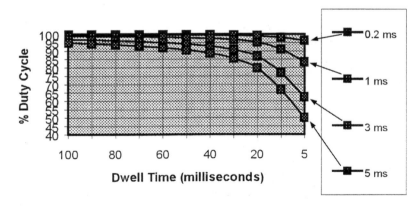

FIGURE 12.9 Measurement duty cycle as a function of dwell time with varying scanning/settling times.

is probably desirable to carry out multiple sweeps with longer dwell times (typically 50 msec) to get the best detection limits. So if an integration time of 1 sec is used for each element, this would translate into 20 sweeps of 50-msec dwell time per mass. While 1 sec is long enough to achieve reasonably good detection limits, longer integration times generally have to be used to reach the lowest possible detection limits. This is exemplified in Figure 12.10, which shows detection limit improvement as a function of integration time for $^{238}U^+$. As would be expected there is a fairly predictable improvement in the detection limit as the integration time is increased because more ions are being counted without an increase in the background noise. However, this only holds true up to the point where the pulse-counting detection system becomes saturated and no more ions can be counted. In the case of $^{238}U^+$, it can be seen that this happens round about 25 sec, because there is no obvious improvement in D/L at a higher integration time. So from this data, we can say that there appears to be no real benefit in using longer than a 7-sec integration time. When deciding the length of the integration time in ICP-MS, you have to weigh up the detection limit improvement against the time taken to achieve that improvement. Is it worth spending 25 sec measuring each mass to get 0.02-ppt detection limit, if 0.03 ppt can be achieved using a 7-sec integration time? Alternatively, is it worth measuring for 7 sec

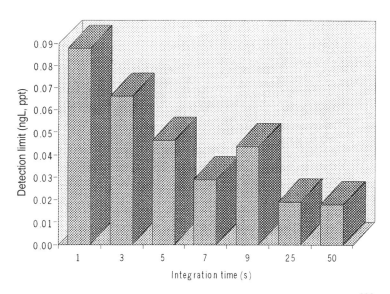

FIGURE 12.10 Plot of detection limit against integration time for $^{238}U^+$. (Courtesy of PerkinElmer Life and Analytical Sciences.)

when 1 sec will only degrade the performance by a factor of 3? It really depends on your data quality objectives.

For some applications such as isotope dilution/ratio studies, high precision is also a very important data quality objective (4). However, to understand what is realistically achievable, we have to be aware of the practical limitations of measuring a signal and counting ions in ICP-MS. Counting statistics tells us that the standard deviation of the ion signal is proportional to the square root of the signal. It follows therefore that the relative standard deviation (RSD) or precision should improve with an increase in the number (N) of ions counted as shown by the following equation:

$$\% \mathrm{RSD} = \frac{\sqrt{N}}{N} \times 100$$

In practice this holds up very well as can be seen in Figure 12.11. In this plot of standard deviation as a function of signal intensity for ^{208}Pb^{+}, the dots represent the theoretical relationship as predicted by counting statistics. It can be seen that the measured standard deviation (bars) follows theory very well up to about 100,000 cps. At that point, additional sources of noise (e.g., sample introduction pulsations/plasma fluctuations) dominate the signal, which lead to poorer standard deviation values (2).

So based on counting statistics, it is logical to assume that the more ions are counted the better the precision will be. To put this in perspective it means that at least 1 million ions need to be counted to achieve an RSD of 0.1%. In practice of course, these kinds of precision values are very difficult

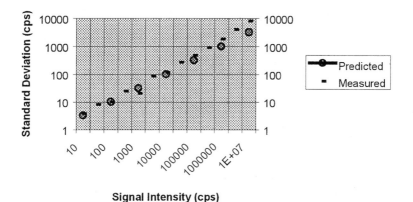

FIGURE 12.11 Comparison of measured standard deviation of a ^{208}Pb^{+} signal against that predicted by counting statistics. (From Ref. 2.)

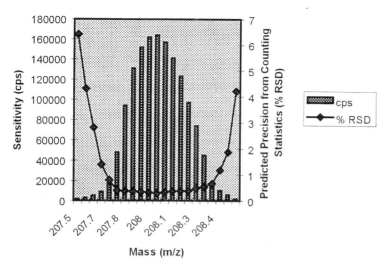

FIGURE 12.12 Comparison of % RSD with signal intensity across the mass profile of a $^{208}Pb^+$ peak. (From Ref. 2.)

to achieve with a scanning quadrupole system, because of the additional sources of noise. If this information is combined with our knowledge of how the quadrupole is scanned, we begin to understand what is required to get the best precision. This is confirmed by the spectral scan in Figure 12.12, which shows the predicted precision at all 20 channels of a 5-ppb $^{208}Pb^+$ peak (2).

This tells us that the best precision is obtained at the channels where the signal is highest, which as we can see are the ones at or near the center of the peak. For this reason, if good precision is a fundamental requirement of your data quality objectives, it is best to use single-point peak hopping with integration times in the order of 5–10 sec. On the other hand, if high precision isotope ratio or isotope dilution work is being done, where analysts would like to achieve precision values approaching counting statistics, then much longer measuring times are required. That is why integration times in the order of 5–10 min are commonly used for determining isotope ratios involving environmental pollutants (5), or clinical metabolism studies (6). For this type of analysis, when two or more isotopes are being measured and ratioed to each other, it follows that the more simultaneous the measurement the better the precision becomes. Therefore the ability to make the measurement as simultaneous as possible is considered more desirable than any other aspect of the measurement. This is supported by the fact that the

TABLE 12.1 Precision of Pb Isotope Ratio Measurement as a Function of Dwell Time Using a Total Integration Time of 5.5 sec

Dwell time (msec)	% RSD $^{207}Pb^+/^{206}Pb^+$	% RSD $^{208}Pb^+/^{206}Pb^+$
2	0.40	0.36
5	0.38	0.36
10	0.23	0.22
25	0.24	0.25
50	0.38	0.33
100	0.41	0.38

Source: From Ref. 7.

best isotope ratio precision data are achieved with time-of-flight (TOF) or multicollector, magnetic sector ICP-MS systems, which are both considered simultaneous in nature. So the best way to approximate simultaneous measurement with a rapid scanning device such as a quadrupole is to use shorter dwell times (but not too short that insufficient ions are counted) and keep the scanning/settling times to an absolute minimum—which results in more sweeps for a given measurement time. This can be seen in Table 12.1, which shows precision of Pb isotope ratios at different dwell times carried out by researchers at the Geological Survey of Israel (7). The data are based on nine replicates of a NIST SRM-981 (75 ppb Pb) solution, using 5.5-sec integration time per isotope.

From these data, the researchers concluded that a dwell time of 10 or 25 msec offered the best isotope ratio precision measurement (quadrupole settling time was fixed at 0.2 msec). They also found that they could achieve slightly better precision by using a 17.5-sec integration time (700 sweeps at 25 msec dwell time) but felt the marginal improvement in precision for nine

TABLE 12.2 Impact of Integration Time on the Overall Analysis Time for Pb Isotope Ratios

Dwell time (msec)	No. of sweeps	Integration time (sec)/mass	%RSD $^{207}Pb^+/^{206}Pb^+$	% RSD $^{207}Pb^+/^{206}Pb^+$	Time for nine replicates (min/sec)
25	220	5.5 sec	0.24	0.25	2 m 29 sec
25	500	12.5 sec	0.21	0.19	6 m 12 sec
25	700	17.5 sec	0.20	0.17	8 m 29 sec

Source: From Ref. 7.

replicates was not worth spending the approximately 3 1/2 times longer analysis time. This can be seen in Table 12.2.

This work shows the benefit of being able to optimize the dwell time, settling time, and the number of sweeps to get the best isotope ratio precision data. They were also very fortunate to be dealing with relatively healthy ion signals for the 3 Pb isotopes, ^{206}Pb, ^{207}Pb, and ^{208}Pb (24.1%, 22.1%, and 52.4% abundance, respectively). If the isotopic signals were dramatically different such as in ^{235}U to ^{238}U (0.72% and 99.2745% abundant, respectively), then the ability to optimize the measurement protocol for individual isotopes becomes of even greater importance to guarantee good precision data.

It is clear that the analytical demands put on ICP-MS are probably higher than any other trace element technique, because it is continually being asked to solve a wide variety of application problems. However, by optimizing the measurement protocol to fit the analytical requirement, ICP-MS has shown that it has the unique capability to carry out rapid trace element analysis, with superb detection limits and good precision on both continuous and transient signals, and still meet the most stringent data quality objectives.

FURTHER READING

1. Integrated MCA Technology in the ELAN ICP-Mass Spectrometer, Application Note TSMS-25, PerkinElmer Instruments, 1993.
2. Denoyer ER. At Spectrosc 1992; 13(3):93–98.
3. Denoyer ER, Lu Q.H. At Spectrosc 1993; 14(6):162–169.
4. Catterick T, Handley H, Merson S. At Spectrosc 1995; 16(10):229–234.
5. Hinners TA, Heithmar EM, Spittler TM, Henshaw JM. Anal Chem 1987; 59:2658–2662.
6. Janghorbani M, Ting BTG, Lynch NE. Microchem Acta 1989; 3:315–328.
7. Halicz L, Erel Y, Veron A. At Spectrosc 1996; 17(5):186–189.

13

Methods of Quantitation

There are many different ways to carry out trace element analysis by ICP-MS depending on your data quality objectives. Such is the flexibility of the technique that it allows detection from sub-parts per thousand up to high parts per million levels using a wide variety of calibration methods from full quantitative and semiquantitative analysis to one of the very powerful isotope ratioing techniques. This chapter takes a look at the most important quantitation methods available in ICP-MS.

This ability of ICP-MS to carry out isotopic measurements allows the technique to carry out quantitation methods, which are not available to any other trace element technique. They include:

Quantitative analysis
Semiquantitative routines
Isotope dilution
Isotope ratio
Internal standardization

Each of these techniques offers varying degrees of accuracy and precision, so it is important to understand their strengths and weaknesses in order to know which one will best meet the data quality objectives. Let us look at each of these in greater detail.

QUANTITATIVE ANALYSIS

As in other maturer trace element techniques like AA and ICP-OES, quantitative analysis in ICP-MS is the fundamental tool used to determine analyte concentrations in unknown samples. In this mode of operation, the instrument is calibrated by measuring the intensity for all elements of interest in a number of known calibration standards that represent a range of concentrations likely to be encountered in your unknown samples. When the full range of calibration standards and blank have been run, the software creates a calibration curve of the measured intensity vs. concentration for each element in the standard solutions. Once calibration data are acquired, the unknown samples are analyzed by plotting the intensity of the elements of interest against the respective calibration curves. The software then calculates the concentrations for the analytes in the unknown samples.

This type of calibration is often called external standardization and is usually used when there is very little difference between the matrix components in the standards and the samples. However, when it is difficult to closely match the matrix of the standards with the samples, external standardization can produce erroneous results because matrix-induced interferences will change analyte sensitivity based on the amount of matrix present in the standards and the samples. When this occurs, better accuracy is achieved by using the method of standard additions or a similar approach called addition calibration. Let us look at these three variations of quantitative analysis to see how they differ.

External Standardization

As explained earlier, this involves measuring a blank solution followed by a set of standard solutions to create a calibration curve over the anticipated concentration range. Typically, a blank and up to three standards containing different analyte concentrations are run. Increasing the number of points on the calibration curve by increasing the number of standards may improve accuracy in circumstances where the calibration range is very broad. However, it is seldom necessary to run a calibration with more than five standards. After the standards have been measured, the unknown samples are analyzed and their analyte intensities are read against the calibration curve. Over extended analysis times, it is common practice to update the calibration curve by either recalibrating the instrument with a full set of standards or by running one midpoint standard. The following protocol summarizes a typical calibration using external standardization:

1. Blank>
2. Std. 1>
3. Std. 2>
4. Std. 3>
5. Sample 1>
6. Sample 2>
7. Sample....n
8. Recalibrate
9. Sample $n + 1$ etc.

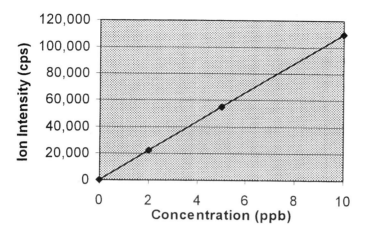

FIGURE **13.1** A simple linear regression calibration curve.

This can be seen more clearly in Figure 13.1, which shows a typical calibration curve using a blank and three standards of 2, 5, and 10 ppb. This calibration curve shows a simple "linear regression," but usually, other modes of calibration are also available like "weighted linear" to emphasize measurements at the low-concentration region of the curve and "linear through zero," where the linear regression is forced through zero.

It should be emphasized that the graph represents a single element calibration. However, because ICP-MS is usually used for multielement analysis, multielement standards are typically used to generate calibration data. For that reason, it is absolutely essential to use multielement standards that have been manufactured specifically for ICP-MS. Single-element AA standards are not suitable because they have only been certified for the analyte element and not for any others. The purity of the standard cannot be guaranteed for any other element and, as result, cannot be used to make up multielement standards for use with ICP-MS. For the same reason, ICP-OES multielement standards are not advisable either because they are only certified for a group of elements and could contain other elements at higher levels, which will affect the ICP-MS multielement calibration.

Standard Additions

This mode of calibration provides an effective way to minimize sample-specific matrix effects by spiking samples with known concentrations of analytes (1,2). In standard addition calibration, the intensity of a blank solution is first measured. Next, the sample solution is "spiked" with known concentrations of each element to be determined. The instrument measures the response for the spiked samples and creates a calibration curve for each element for which a spike has been added. The calibration curve is a plot of the blank-subtracted intensity of each spiked element against its concentration value. After creating the calibration curve, the unspiked sample solutions are then analyzed and compared to the calibration curve. Based on the slope of the calibration curve and where it intercepts the x-axis, the instrument software determines the unspiked concentration of the analytes in the unknown samples. This can be seen in Figure 13.2, which shows a calibration of the sample intensity plus the sample spiked with 2 and 5 ppb of the analyte. The concentration of the sample is where the calibration line intercepts the negative side of the x-axis.

The following protocol summarizes a typical calibration using the method of standard additions.

    1.   Blank>
    2.   Spiked sample 1 (spike conc. 1)>
    3.   Spiked sample 1 (spike conc. 2)>
    4.   Unspiked sample 1>
    5.   Blank>
    6.   Spiked sample 2 (spike conc. 1)>
    7.   Spiked sample 2 (spike conc. 2)>
    8.   Unspiked sample 2>
    9.   Blank>
   10.   Etc.

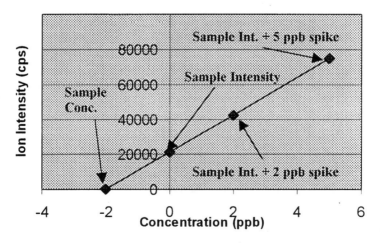

FIGURE 13.2 A typical "Method of Additions" calibration curve.

Addition Calibration

Unfortunately, with the method of standard additions, each and every sample has to be spiked with all the analytes of interest, which becomes extremely labor-intensive when many samples have to be analyzed. For this reason, a variation of standard additions called "addition calibration" is more widely used in ICP-MS. However, this method can only be used when all the samples have a similar matrix. It uses the same principle as standard additions, but only the first (or representative) sample is spiked with known concentrations of analytes and then analyzes the rest of the sample batch against the calibration assuming all samples have a similar matrix to the first one. The following protocol summarizes a typical calibration using the method of addition calibration.

1. Blank>
2. Spiked sample 1 (spike cont. 1)>
3. Spiked sample 1 (spike cont. 2)>
4. Unspiked sample 1>
5. Unspiked sample 2>
6. Unspiked sample 3>
7. Etc.

SEMIQUANTITATIVE ANALYSIS

If your data quality objectives for accuracy and precision are less stringent, ICP-MS offers a very rapid semiquantitative mode of analysis. This technique enables you to automatically determine the concentrations of approximately 80 elements in an unknown sample, without the need for calibration standards (3,4). There are slight variations in the way different instruments approach semiquantitative analysis, but the general principle is to measure the entire mass spectrum without specifying individual elements or masses. It relies on the principle that each element's natural isotopic abundance is fixed. By measuring the intensity of all their isotopes, correcting for common spectral interferences, including molecular, polyatomic, and isobaric species, and applying heuristic, knowledge-driven routines in combination with numerical calculations, a positive or a negative confirmation can be made for each element present in the sample. Then by comparing the corrected intensities against a stored isotopic response table, a good semiquantitative approximation of the sample components can be made.

Semiquant, as it is often called, is an excellent approach to rapidly characterize unknown samples. Once the sample has been characterized, you can choose to either update the response table with your own standard solutions to improve analytical accuracy or switch to the quantitative analysis mode to focus on specific elements and determine their concentrations with even greater accuracy and precision. While a semiquantitative determination can be performed without using a series of standards, the use of a small number of standards is highly recommended for improved accuracy across the full mass range. Unlike traditional quantitative analysis in which you analyze standards for all the elements you want to determine, semiquant calibration is achieved using just a few elements distributed across the mass range. This calibration process, shown more clearly in Figure 13.3, is used to update the reference response curve data that correlate measured ion intensities to the concentrations of elements in a solution. During calibration, these response data are adjusted to account for changes in the instrument's sensitivity due to variations in the sample matrix.

This process is often called semiquantitative analysis using external calibration, and like traditional quantitative analysis using external standardization, it works extremely well for samples which all have a similar matrix. However, if you are analyzing samples containing widely different concentrations of matrix components, external calibration does not work very well because of matrix-induced suppression effects on the analyte signal. If this is the case, semiquant using a variation of standard addition calibration should be used. Similar to standard addition calibration used in quantitative analysis, this procedure involves adding known quantities of specific elements to every unknown sample before measurement. The major difference with semiquant is that the elements you add must not already be present in significant quantities

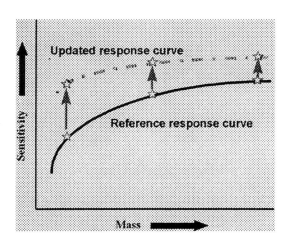

FIGURE 13.3 In semiquantitative analysis, a small group of elements is used to update the reference response curve to improve the accuracy as the sample matrix changes.

in the unknown samples because they are being used to update the stored reference response curve. As with external calibration, the semiquant software then adjusts the stored response data for all the remaining analytes relative to the calibration elements. This procedure works very well but tends to be very labor-intensive because the calibration standards have to be added to every unknown sample.

ISOTOPE DILUTION

Although quantitative and semiquantitative analysis methods are suitable for the majority of applications, there are other calibration methods available, depending on your analytical requirements. For example, if your application requires even greater accuracy and precision, the "isotope dilution" technique may offer some benefits. Isotope dilution is an absolute means of quantitation based on altering the natural abundance of two isotopes of an element by adding a known amount of one of the isotopes and is considered one of the most accurate and precise approaches to elemental analysis (5–8).

For this reason, a prerequisite of isotope dilution is that the element must have at least two stable isotopes. The principle works by spiking a known weight of an enriched stable isotope into your sample solution. By knowing the natural abundance of the two isotopes being measured, the abundance of the spiked enriched isotopes, the weight of the spike, and the weight of the sample, the original trace element concentration can be determined by using the following equation:

$$C = \frac{[\text{Aspike} - (R \times \text{Bspike})] \times \text{Wspike}}{[R \times (\text{Bsample} - \text{Asample})] \times \text{Wsample}}$$

where

C = concentration of trace element
Aspike = % of higher abundance isotope in spiked enriched isotope
Bspike = % of lower abundance isotope in spiked enriched isotope
Wspike = weight of spiked enriched isotope
R = ratio of the % of higher abundance isotope to lower abundance isotope in the spiked sample
Bsample = % of higher natural abundance isotope in sample
Asample = % of lower natural abundance isotope in sample
Wsample = weight of sample

This might sound complicated, but in practice, it is relatively straightforward. This is exemplified in Figure 13.4, which shows an isotope dilution

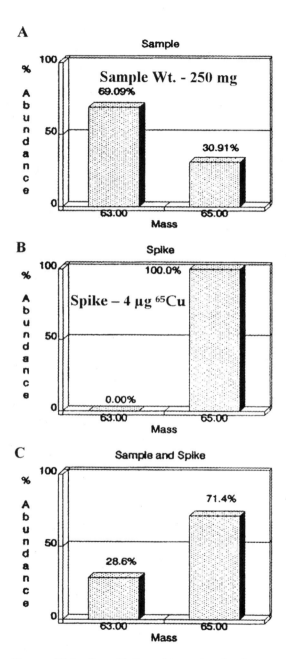

FIGURE 13.4 Quantitation of trace levels of copper in a sample of SRM orchard leaves using isotope dilution methodology. (From Ref. 9.)

method for the determination of copper in a 250-mg sample of orchard leaves, using the two copper isotopes ^{63}Cu and ^{65}Cu.

In Figure 13.4A, it can be seen that the natural abundance of the two isotopes are 69.09% and 30. 91% for ^{63}Cu and ^{65}Cu, respectively. Figure 13.4B shows that 4 μg of an enriched isotope of 100% ^{65}Cu (and 0 % ^{63}Cu) is spiked into the sample, which now produces a spiked sample containing 71.4% of ^{65}Cu and 28.6% of ^{63}Cu, as seen in Figure 13.4C (9). If we plug these data into the equation above, we obtain:

$$ C = \frac{[100 - (71.4/28.6 \times 0)] \times 4 \ \mu g}{[(71.4/28.6 \times 69.09) - 30.91] \times 0.25 \ g} $$

$$ C = 400/35.45 = 11.3 \ \mu g/g $$

The major benefit of the isotope dilution technique is that it provides measurements that are extremely accurate because you are measuring the concentration of the isotopes in the same solution as your unknown sample and not in a separate external calibration solution. In addition, because it is a ratioing technique, the loss of solution during the sample preparation stage has no influence on the accuracy of the result. The technique is also extremely precise because using a simultaneous detection system like a magnetic sector multicollector or a simultaneous ion sampling device like a TOF ICP-MS, the results are based on measuring the two isotope solutions at the same time, which compensates for imprecision of the signal due to sources of sample introduction-related noise, such as plasma instability, peristaltic pump pulsations, and nebulization fluctuations. Even using a scanning mass analyzer like a quadrupole, the measurement protocol can be optimized to scan very rapidly between the two isotopes and achieve very good precision. However, isotope dilution has some limitations, which makes it only suitable for certain applications. These limitations include the following.

    The element you are determining must have more than one isotope because calculations are based on the ratio of one isotope to another isotope of the same element—this makes it unsuitable for approximately 15 elements that can be determined by ICP-MS.
    It requires certified enriched isotopic standards, which can be very expensive, especially those that are significantly different from the normal isotopic abundance of the element.
    It compensates for interferences due to signal enhancement or suppression, but does not compensate for spectral interferences. For this reason, an external blank solution must always be run.

ISOTOPE RATIOS

The ability of ICP-MS to determine individual isotopes also makes it suitable for another isotopic measurement technique called "isotope ratio" analysis. The ratio of two or more isotopes in a sample can be used to generate very useful information including an indication of the age of a geological formation, a better understanding of animal metabolism, and also help to identify sources of environmental contamination (10–14). Similar to isotope dilution, isotope ratio analysis uses the principle of measuring the exact ratio of two isotopes of an element in the sample. With this approach, the isotope of interest is typically compared to a reference isotope of the same element. For example, you might want to compare the concentration of ^{204}Pb to the concentration of ^{206}Pb. Alternatively, the requirement might be to compare one isotope to all remaining reference isotopes of an element, like the ratio of ^{204}Pb to ^{206}Pb, ^{207}Pb, and ^{208}Pb. The ratio is then expressed in the following manner:

$$\text{Isotope ratio} = \frac{\text{Intensity of isotope of interest}}{\text{Intensity of reference isotope}}$$

Since this ratio can be calculated from within a single sample measurement, classic external calibration is not normally required. However, if there is a large difference between the concentrations of the two isotopes, it is recommended to run a standard of known isotopic composition. This is done to verify that the higher-concentration isotope is not suppressing the signal of the lower-concentration isotope and biasing the results. This effect called mass discrimination is less of a problem if the isotopes are relatively close in concentration like ^{107}Ag to ^{109}Ag, which are 51.839% and 48.161% abundant, respectively. However, it can be an issue if there is a significant difference in their concentration values such as ^{235}U to ^{238}U, which are 0.72% and 99.275% abundant, respectively. Mass discrimination effects can be reduced by running an external reference standard of known isotopic concentration, comparing the isotope ratio with the theoretical value, and then mathematically compensating for the difference.

INTERNAL STANDARDIZATION

Another method of standardization commonly employed in ICP-MS is called "internal standardization." It is not considered an absolute calibration technique, but instead used to correct for changes in analyte sensitivity caused by variations in the concentration and type of matrix components found in the sample. An internal standard is a nonanalyte isotope that is added to the blank solution, standards, and samples before analysis. It is typical to add three or four internal standard elements to the samples to cover the analyte elements of interest. The software adjusts the analyte concentration in the unknown samples by comparing the intensity values of the internal standard intensities in the unknown sample to those in the calibration standards.

The implementation of internal standardization varies according to the analytical technique which is being used. For quantitative analysis, the internal standard elements are selected based on the similarity of their ionization characteristics to the analyte elements. Each internal standard is bracketed with a group of analytes. The software then assumes that the intensities of all elements within a group are affected in a similar manner by the matrix. Changes in the ratios of the internal standard intensities are then used to correct the analyte concentrations in the unknown samples.

For semiquantitative analysis that uses a stored response table, the purpose of the internal standard is similar, but a little different in implementation to quantitative analysis. A semiquant internal standard is used to continuously compensate for instrument drift or matrix-induced suppression over a defined mass range. If a single internal standard is used, all the masses selected for the determination are updated by the same amount based on the intensity of the internal standard. If more than one internal standard is used, which is recommended for measurements over a wide mass range, the software interpolates the intensity values based on the distance in mass between the analyte and the nearest internal standard element.

It is worth emphasizing that if you do not want to compare your intensity values to a calibration graph, most instruments allow you to report raw data. This enables you to analyze your data using external data-processing routines, to selectively apply a minimum set of ICP-MS data-processing methods, or just to view the raw data file before reprocessing it. The availability of raw data is primarily intended for use in nonroutine applications like chromatography separation techniques and laser sampling devices that produce a time-resolved transient peak or by users whose sample set requires data processing using algorithms other than those supplied by the instrument software.

FURTHER READING

1. Beauchemin D, McLaren JW, Mykytiuk AP, Berman SS. Anal Chem 1987; 59:778.
2. Pruszkowski E, Neubauer K, Thomas R. At Spectr 1998; 19(4):111–115.
3. Broadhead M, Broadhead R, Hager JW. At Spectr 1990; 11(6):205–209.
4. Denoyer E. J Anal At Spectrom 1992; 7:1187.
5. McLaren JW, Beauchemin D, Berman SS. Anal Chem 1987; 59:610.
6. Longerich H. At Spectr 1989; 10(4):112–115.

7. Stroh A. At Spectr 1993; 14(5):141–143.
8. Catterick T, Handley H, Merson S. At Spectr 1995; 16(10):229–234.
9. Multi-Elemental Isotope Dilution Using the Elan ICP-MS Elemental Analyzer; ICP-MS Technical Summary TSMS-1, PerkinElmer Instruments, 1985.
10. Ting BTG, Janghorbani M. Anal Chem 1986; 58:1334.
11. Janghorbani M, Ting BTG, Lynch NE. Microchem Acta 1989; 3:315–328.
12. Hinners TA, Heithmar EM, Spittler TM, Henshaw JM. Anal Chem 1987; 59: 2658–2662.
13. Halicz L, Erel Y, Veron A. At Spectr 1996; 17(5):186–189.
14. Chaudhary-Webb M, Paschal DC, Elliott WC, Hopkins HP, Ghazi AM, Ting BC, Romieu I. At Spectr 1998; 19(5):156.

14

Review of Interferences

Now that we have covered the major instrumental components of an ICP mass spectrometer, let us turn our attention to the technique's most common interferences and what methods are used to compensate for them. Although interferences are reasonably well understood in ICP-MS, it can often be difficult and time-consuming to compensate for them, particularly in complex sample matrices. Having prior knowledge of the interferences associated with a particular set of samples will often dictate the sample preparation steps and the instrumental methodology used to analyze them.

Interferences in ICP-MS are generally classified into three major groups—spectral-, matrix-, and physical-based interferences. Each of them has the potential to be problematic in its own right, but modern instrumentation and good software combined with optimized analytical methodologies have minimized their negative impact on trace element determinations by ICP-MS. Let us take a look at these interferences in greater detail and describe the different approaches used to compensate for them.

SPECTRAL INTERFERENCES

Spectral overlaps are probably the most serious types of interferences seen in ICP-MS. The most common type is known as a polyatomic or molecular spectral interference, which is produced by the combination of two or more atomic ions. They are caused by a variety of factors but are usually associated with either the plasma/nebulizer gas used, matrix components in the solvent/sample, other elements in the sample, or entrained oxygen/nitrogen from the surrounding air. For example, in the argon plasma, spectral overlaps caused by argon ions and combinations of argon ions with other species are very common. The most abundant isotope of argon is at mass 40, which dramatically interferes with the most abundant isotope of calcium at mass 40, whereas the combination of argon and oxygen in an aqueous sample generates the $^{40}Ar^{16}O^+$ interference, which has a significant impact on the major isotope of Fe at mass 56. The complexity of these kinds of spectral problems can be seen in Figure 14.1, which shows a mass spectrum of deionized water from mass 40 to mass 90.

Additionally, argon can also form polyatomic interferences with elements found in the acids used to dissolve the sample. For example, in a hydrochloric acid medium, $^{40}Ar^+$ combines with the most abundant chlorine isotope at 35 amu to form $^{40}Ar^{35}Cl^+$, which interferes with the only isotope of arsenic at mass 75, while in an organic solvent matrix, argon and carbon combine to form $^{40}Ar^{12}C^+$, which interferes with $^{52}Cr^+$, the most abundant isotope of chromium. Sometimes, matrix/solvent species need no help from argon ions and combine to form spectral interferences of their own. A good example is in a sample that contains sulfuric acid. The dominant sulfur isotope, $^{32}S^+$, combines with two oxygen ions to form a $^{32}S^{16}O^{16}O^+$ molecular ion, which interferes with the major isotope of Zn at mass 64. In the analysis of samples containing high concentrations of sodium, such as seawater, the most abundant isotope of Cu at mass 63 cannot be used because of interference from the $^{40}Ar^{23}Na^+$ molecular ion. There are many more examples of these kinds of polyatomic and molecular interferences, which have been comprehensively reviewed in the literature (1). Table 14.1 represents some of the most common ones seen in ICP-MS.

Oxides, Hydroxides, Hydrides, and Doubly Charged Species

Another type of spectral interference is produced by elements in the sample combining with H^+, $^{16}O^+$, or $^{16}OH^+$ (either from water or air) to form

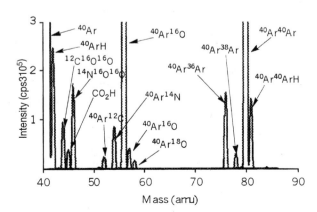

FIGURE 14.1   ICP mass spectrum of deionized water from mass 40 to mass 90.

TABLE 14.1 Some Common Plasma/Matrix/Solvent-Related Polyatomic Spectral Interferences Seen in ICP-MS

Element/isotope	Matrix/solvent	Interference
$^{39}K^+$	H_2O	$^{38}ArH^+$
$^{40}Ca^+$	H_2O	$^{40}Ar^+$
$^{56}Fe^+$	H_2O	$^{40}Ar^{16}O^+$
$^{80}Se^+$	H_2O	$^{40}Ar^{40}Ar^+$
$^{51}V^+$	HCl	$^{35}Cl^{16}O^+$
$^{75}As^+$	HCl	$^{40}Ar^{35}Cl^+$
$^{28}Si^+$	HNO_3	$^{14}N^{14}N^+$
$^{44}Ca^+$	HNO_3	$^{14}N^{14}N^{16}O^+$
$^{55}Mn^+$	HNO_3	$^{40}Ar^{15}N^+$
$^{48}Ti^+$	H_2SO_4	$^{32}S^{16}O^+$
$^{52}Cr^+$	H_2SO_4	$^{34}S^{18}O^+$
$^{64}Zn^+$	H_2SO_4	$^{32}S^{16}O^{16}O^+$
$^{63}Cu^+$	H_3PO_4	$^{31}P^{16}O^{16}O^+$
$^{24}Mg^+$	Organics	$^{12}C^{12}C^+$
$^{52}Cr^+$	Organics	$^{40}Ar^{12}C^+$
$^{65}Cu^+$	Minerals	$^{48}Ca^{16}OH^+$
$^{64}Zn^+$	Minerals	$^{48}Ca^{16}O^+$
$^{63}Cu^+$	Seawater	$^{40}Ar^{23}Na^+$

molecular hydrides ($+H^+$), oxides ($+^{16}O^+$), and hydroxides ($+^{16}OH^+$) ions, which occur at 1, 16, and 17 mass units, respectively, higher than the element's mass (2). These interferences are typically produced in the cooler zones of the plasma, immediately before the interface region. They are usually more serious when rare earth or refractory-type elements are present in the sample, because many of them readily form molecular species (particularly oxides), which create spectral overlap problems on other elements in the same group. If the oxide species is mainly derived from entrained air around the plasma, it can be reduced by either using an elongated outer tube to the torch, or using a metal shield between the plasma and the RF coil.

Associated with oxide-based spectral overlaps are doubly charged spectral interferences. These are species that are formed when an ion is generated with a double positive charge as opposed to a normal single charge and produces an isotopic peak at half its mass. Like the formation of oxides, the level of doubly charged species is related to the ionization conditions in the plasma and can usually be minimized by careful optimization of the nebulizer gas flow, RF power, and sampling position within the plasma. It can also be impacted by the severity of the secondary discharge present at the interface (3), which was described in greater detail in Chapter 5. Table 14.2 shows a selected group of elements, which readily form oxides/hydroxides/hydrides

TABLE 14.2   Some Elements That Readily Form
Oxides, Hydroxides, Hydrides, and Doubly Charged
Species in the Plasma, Together with the Analytes
Affected by the Interference

Oxide, hydroxide, hydride, doubly charged species	Analyte affected by interference
$^{40}Ca^{16}O^+$	$^{56}Fe^+$
$^{48}Ti^{16}O^+$	$^{64}Zn^+$
$^{98}Mo^{16}O^+$	$^{114}Cd^+$
$^{138}Ba^{16}O^+$	$^{154}Sm^+$, $^{154}Gd^+$
$^{139}La^{16}O^+$	$^{155}Gd^+$
$^{140}Ce^{16}O^+$	$^{156}Gd^+$, $^{156}Dy^+$
$^{40}Ca^{16}OH^+$	$^{57}Fe^+$
$^{31}P^{18}O^{16}OH^+$	$^{66}Zn^+$
$^{79}BrH^+$	$^{80}Se^+$
$^{31}P^{16}O_2H^+$	$^{64}Zn^+$
$^{138}Ba^{2+}$	$^{69}Ga^+$
$^{139}La^{2+}$	$^{69}Ga^+$
$^{140}Ce^{2+}$	$^{70}Ge^+$, $^{70}Zn^+$

and doubly charged species, together with the analytes that are affected by them.

Isobaric Interferences

The final classification of spectral interferences is called isobaric overlaps, produced mainly by different isotopes of other elements in the sample creating spectral interferences at the same mass as the analyte. For example, vanadium has two isotopes at 50 and 51 amu. However, mass 50 is the only practical isotope to use in the presence of a chloride matrix, because of the large contribution from the $^{16}O^{35}Cl^+$ interference at mass 51. Unfortunately, mass 50 amu, which is only 0.25% abundant, also coincides with isotopes of titanium and chromium, which are 5.4% and 4.3% abundant, respectively. This makes the determination of vanadium in the presence of titanium and chromium very difficult unless mathematical corrections are made. Table 14.3—relative abundances of the isotopes—shows all the possible naturally occurring isobaric spectral overlaps in ICP-MS (4).

Ways to Compensate for Spectral Interferences

Let us now look at the different approaches used to compensate for spectral interferences. One of the very first ways used to get around severe matrix-derived spectral interferences was to remove the matrix somehow. In the early days this involved precipitating the matrix with a complexing agent and then filtering off the precipitate. However, more recently this has been carried out by automated matrix removal/analyte preconcentration techniques using chromatography-type equipment. In fact, this is the preferred method for carrying out trace metal determinations in seawater, because of the matrix and spectral problems associated with such high concentrations of sodium and magnesium chloride (5).

Mathematical Correction Equations

Another method that has been successfully used to compensate for isobaric interferences and some less severe polyatomic overlaps (when no alternative isotopes are available for quantitation) is to use mathematical interference correction equations. Similar to interelement corrections (IECs) in ICP-OES, this method works on the principle of measuring the intensity of the interfering isotope or interfering species at another mass, which is ideally free of any interferences. A correction is then applied by knowing the ratio of the intensity of the interfering species at the analyte mass to its intensity at the alternate mass. Let us take a look at a "real-world" example to exemplify this type of correction. The most sensitive isotope for cadmium is at mass 114. However, there is also a minor isotope of tin at mass 114. This means that if there is any tin in the sample, quantitation using $^{114}Cd^+$ can only be carried out if a correction is made for $^{114}Sn^+$. Fortunately, Sn has a total of 10 isotopes, which means that there is probably going to be at least one of them free of a spectral interference. Therefore by measuring the intensity of Sn at one of its most abundant isotopes (typically $^{118}Sn^+$) and ratioing it to $^{114}Sn^+$, a correction is made in the method software—in the following manner:

Total counts at mass 114 = $^{114}Cd^+$ + $^{114}Sn^+$
Therefore $^{114}Cd^+$ = Total counts at mass 114 − $^{114}Sn^+$

To find out the contribution from $^{114}Sn^+$, it is measured at the interference free isotope of $^{118}Sn^+$ and a correction of the ratio of $^{114}Sn^+/^{118}Sn^+$ is applied:

Which means $^{114}Cd^+$ = Counts at mass 114 − ($^{114}Sn^+/^{118}Sn^+$) × ($^{118}Sn^+$)

Now the ratio ($^{114}Sn^+/^{118}Sn^+$) is the ratio of the natural abundances of these two isotopes (065%/24.23%) and is always constant

Therefore $^{114}Cd^+$ = mass 114 − (0.65%/24.23%) × ($^{118}Sn^+$)
or $^{114}Cd^+$ = mass 114 − (0.0268) × ($^{118}Sn^+$)

An interference correction for $^{114}Cd^+$ would then be entered in the software as:

$-(0.0268)*(^{118}Sn^+)$

TABLE 14.3 Relative Isotopic Abundances of the Naturally Occurring Elements, Showing All the Potential Isobaric Interferences

Isotope		Relative abundance of the natural isotopes				
		%		%		%
1	H	98.985				
2	H	0.015				
3			He	0.000137		
4			He	99.999863		
5						
6					Li	7.5
7					Li	82.5
8						
9	Be	100				
10			B	19.9		
11			B	80.1		
12					C	98.9
13					C	1.10
14	N	99.643				
15	N	0.365				
16			O	99.762		
17			O	0.038		
18			O	0.200		
19					F	100
20	Ne	90.48				
21	Ne	0.27				
22	Ne	9.25				
23			Na	100		
24					Mg	78.99
25					Mg	10.00
26					Mg	11.01
27	Al	100				
28			Si	92.23		
29			Si	4.67		
30			Si	3.10		
31					P	100
32	S	95.02				
33	S	0.75				
34	S	4.21				
35			Cl	75.77		
36	S	0.02			Ar	0.337
37			Cl	24.23		
38					Ar	0.063
39	K	93.2581				
40	K	0.0117	Ca	96.941	Ar	99.600

Isotope	Relative abundance of the natural isotopes					
	%		%		%	
41	K	6.7302				
42			Ca	0.647		
43			Ca	0.135		
44			Ca	2.086		
45					Sc	100
46	Tl	8.0	Ca	0.004		
47	Tl	7.3				
48	Tl	73.8	Ca	0.187		
49	Tl	5.5				
50	Tl	5.4	V	0.250	Cr	4.345
51			V	96.750		
52					Cr	83.789
53					Cr	9.501
54	Fe	5.8			Cr	2.365
55			Mn	100		
56	Fe	91.72				
57	Fe	2.2				
58	Fe	0.28			Ni	68.077
59			Co	100	Ni	26.233
60					Ni	1.140
61					Ni	3.634
62						
63	Cu	69.17				
64			Zn	48.6	Ni	0.926
65	Cu	30.83				
66			Zn	27.9		
67			Zn	4.1		
68			Zn	18.8		
69					Ga	60.108
70	Ga	21.23	Zn	0.6		
71					Ge	39.892
72	Ge	27.66				
73	Ge	7.73				
74	Ge	35.94	Se	0.89		
75					As	100
76	Ge	7.44	Se	9.36		
77			Se	7.63		
78	Kr	0.35	Se	23.68		
79					Br	50.69
80	Kr	2.25	Se	49.61		
81					Br	49.31

TABLE 14.3 Continued

Isotope		Relative abundance of the natural isotopes %		%		%
82	Kr	11.6	Se	8.73		
83	Kr	11.5				
84	Kr	57.0	Sn	0.56		
85					Pb	72.165
86	Kr	17.3	Sr	9.86		
87			Sr	7.00	Rb	27.835
88			Sr	82.58		
89					Y	100
90	Zr	51.45				
91	Zr	11.22				
92	Zr	17.15	Mo	14.84		
93					Nb	100
94	Zr	17.38	Mo	9.25		
95			Mo	15.92		
96	Zr	2.80	Mo	16.68	Ru	5.52
97			Mo	9.55		
98			Mo	24.13	Ru	1.88
99					Ru	12.7
100			Mo	9.63	Ru	12.6
101					Ru	17.0
102	Pd	1.02			Ru	31.6
103			Rh	100		
104	Pd	11.14				
105	Pd	22.33				
106	Pd	27.33	Cd	1.25		
107					Ag	51.839
108	Pd	26.46	Cd	0.89		
109					Ag	48.161
110	Pd	11.72	Cd	12.49		
111			Cd	12.80		
112	Sn	0.97	Cd	24.13		
113			Cd	12.22	In	4.3
114	Sn	0.65	Cd	28.73		
115	Sn	0.34			In	95.7
116	Sn	14.53	Cd	7.49		
117	Sn	7.68				
118	Sn	24.23				
119	Sn	8.59				
120	Sn	32.59	Te	0.96		
121					Sb	57.36

Isotope	Relative abundance of the natural isotopes					
		%		%		%
122	Sn	4.63	Te	2.603		
123			Te	0.908	Sb	32.64
124	Sn	5.79	Te	4.816	Xe	0.10
125			Te	7.139		
126			Te	18.95	Xe	0.09
127	I	100				
128			Te	31.69	Xe	1.91
129					Xe	26.4
130	Be	0.106	Te	33.80	Xe	4.1
131					Xe	21.2
132	Ba	0.101			Xe	26.9
133			Cs	100		
134	Ba	2.417			Xe	10.4
135	Ba	6.592				
136	Ba	7.854	Ce	0.19	Xe	8.9
137	Ba	11.23				
138	Ba	71.70	Ce	0.25	La	0.0902
139					La	99.9098
140			Ce	88.48		
141					Pr	100
142	Nd	27.13	Ce	11.08		
143	Nd	12.18				
144	Nd	23.80	Sm	3.1		
145	Nd	8.30				
146	Nd	17.19				
147			Sm	15.0		
148	Nd	5.76	Sm	11.3		
149			Sm	13.8		
150	Nd	5.64	Sm	7.4		
151					Eu	47.8
152	Gd	0.20	Sm	26.7		
153					Eu	52.2
154	Gd	2.18	Sm	22.7		
155	Gd	14.80				
156	Gd	20.47	Dy	0.06		
157	Gd	15.65				
158	Gd	24.84	Dy	0.10		
159					Tb	100
160	Gd	21.86	Dy	2.34		
161			Dy	18.9		

TABLE 14.3   Continued

Isotope	Relative abundance of the natural isotopes					
		%		%		%
162	Er	0.14	Dy	25.5		
163			Dy	24.9		
164	Er	1.61	Dy	28.2		
165					Ho	100
166	Er	33.6				
167	Er	22.95				
168	Er	26.8	Yb	0.13		
169					Tm	100
170	Er	14.9	Yb	3.05		
171			Yb	14.3		
172			Yb	21.9		
173			Yb	16.2		
174			Yb	31.8	Hf	0.162
175	Lu	97.41				
176	Lu	2.59	Yb	12.7	Hf	5.206
177					Hf	18.806
178					Hf	27.297
179					Hf	13.629
180	Ta	0.012	W	0.13	Hf	35.100
181	Ta	99.988				
182			W	26.3		
183			W	14.3		
184	Os	0.02	W	30.67		
185					Re	37.40
186	Os	1.58	W	28.6		
187	Os	1.6			Re	62.60
188	Os	13.3				
189	Os	16.1				
190	Os	26.4			Pt	0.01
191			Ir	37.3		
192	Os	41.0			Pt	0.79
193			Ir	62.7		
194					Pt	32.9
195					Pt	33.8
196	Hg	0.15			Pt	25.3
197			Au	100		
198	Hg	9.97			Pt	7.2
199	Hg	16.87				
200	Hg	23.10				
201	Hg	13.18				

Isotope	Relative abundance of the natural isotopes					
		%		%		%
202	Hg	29.86				
203					Tl	29.524
204	Hg	6.87	Pb	1.4		
205					Tl	70.476
206			Pb	24.1		
207			Pb	22.1		
208			Pb	52.4		
209	Bi	100				
210						
211						
212						
213						
214						
215						
216						
217						
218						
219						
220						
221						
222						
223						
224						
225						
226						
227						
228						
229						
230						
231	Pa	100				
232	Th	100				
233						
234	U	0.0055				
235	U	0.7200				
236						
237						
238	U	99.2745				

Source: From Ref. 4.

This is a relatively simple example but explains the basic principles of the process. In practice, especially in spectrally complex samples, corrections often have to be made to the isotope being used for the correction—in addition to the analyte mass, which makes the mathematical equation far more complex.

This approach can also be used for some less severe polyatomic-type spectral interferences. For example, in the determination of V at mass 51 in diluted brine (typically 1000 ppm NaCl), there is a substantial spectral interference from $^{35}C^{16}lO^+$ at mass 51. By measuring the intensity of the $^{37}C^{16}lO^+$ at mass 53, which is free of any interference, a correction can be applied in a similar way to the previous example.

Cool/Cold Plasma Technology

If the intensity of the interference is large, and analyte intensity is extremely low, mathematical equations are not ideally suited as a correction method. For that reason, alternative approaches have to be considered to compensate for the interference. One such approach, which has helped to reduce some of the severe polyatomic overlaps, is to use cold/cool plasma conditions. This technology, which was reported in the literature in the late 1980s, uses a low-temperature plasma to minimize the formation of certain argon-based polyatomic species (6). Under normal plasma conditions (typically 1000–1400 W RF power and 0.8–1.0 L/min of nebulizer gas flow), argon ions combine with matrix and solvent components to generate problematic spectral interferences such as $^{38}ArH^+$, $^{40}Ar^+$, and $^{40}Ar^{16}O^+$, which impact the detection limits of a small number of elements including K, Ca, and Fe. By using cool plasma conditions (500–800 W RF power and 1.5–1.8 L/min nebulizer gas flow), the ionization conditions in the plasma are changed so that many of these interferences are dramatically reduced. The result is that detection limits for this group of elements are significantly enhanced (7). An example of this improvement is seen in Figure 14.2. It shows a spectral scan of 100 ppt of $^{56}Fe^+$ (its most sensitive isotope) using cool plasma conditions. It can be clearly seen that there is virtually no contribution from $^{40}Ar^{16}O^+$, as indicated by the extremely low background for deionized water, resulting in single figure ppt detection limits for iron. Under normal plasma conditions, the $^{40}Ar^{16}O^+$ intensity is so large that it would completely overlap the $^{56}Fe^+$ peak.

Cool plasma conditions are limited to a small group of elements in aqueous-type solutions that are prone to argon-based spectral interferences. It offers very little benefit for the majority of the other elements, because its ionization temperature is significantly lower than a normal plasma. For this reason, it is not ideally suited for the analysis of complex samples, because of

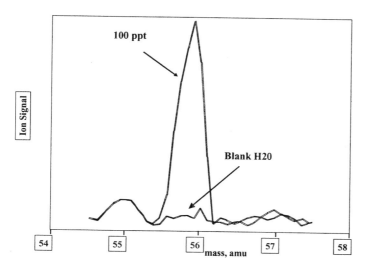

FIGURE 14.2 Spectral scan of 100 ppt ^{56}Fe and deionized water using cool plasma conditions. (From Ref. 8.)

severe signal suppression caused by the matrix. However, it does offer real detection limit improvement for elements with low ionization potential such as sodium and lithium, which benefit from the ionization conditions of the cooler plasma.

Collision/Reaction Cells

The limitations of cool plasmas have led to the development of collision and reaction cells, which utilize ion-molecule collisions and reactions to cleanse the ion beam of harmful polyatomic and molecular interferences, before they enter the mass analyzer. Collision/reaction cells are showing enormous potential to eliminate spectral interferences and make available isotopes that were previously unavailable for quantitation. For example, Figure 14.3 shows a spectral scan of 50 ppt arsenic in 1000 ppm NaCl, together with 1000 ppm NaCl at mass 75, using a dynamic reaction cell with hydrogen/argon mixture as the reaction gas. It can be seen that there is insignificant contribution from the $^{40}Ar^{35}Cl^+$ interference, as indicated by the NaCl baseline. The capability of this type of cell to virtually eliminate the $^{40}Ar^{35}Cl^+$ interference now makes it possible to determine low ppt levels of mono-isotopic $^{75}As^+$ in a high chloride matrix—previously not achievable by conventional interference correction methods (9). A full review of collision/reaction cell technology is given in Chapter 10.

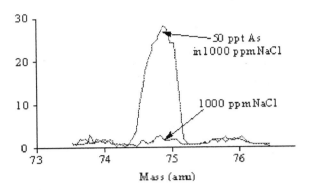

FIGURE 14.3 Reduction of the $^{40}Ar^{35}Cl^+$ interference makes it possible to determine low ppt levels of mono-isotopic $^{75}As^+$ in a high chloride matrix using dynamic reaction cell technology. (From Ref. 9.)

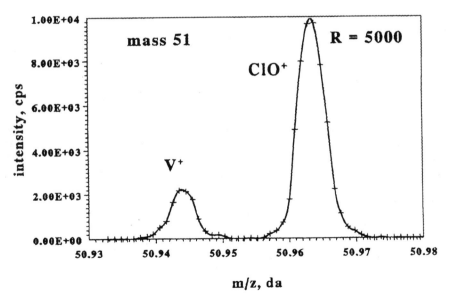

FIGURE 14.4 Separation of $^{51}V^+$ from $^{35}Cl^{16}O^+$ using high resolving power (5000) of a double focusing magnetic sector instrument. (From Ref. 11.)

High-Resolution Mass Analyzers

The best and probably most efficient way to remove spectral overlaps is to resolve them away using a high-resolution mass spectrometer (10). Over the past 10 years, this approach, particularly double focusing magnetic sector mass analyzers, has proved to be invaluable for separating many of the problematic polyatomic and molecular interferences seen in ICP-MS, without the need to use cool plasma conditions or collision/reaction cells. This can be seen in Figure 14.4, which shows a spectral peak for 20 ppb of $^{51}V^+$ resolved from the $^{35}Cl^{16}O^+$ interference in a 0.4 M hydrochloric acid matrix, using a resolution setting of 5000 (11).

However, although their resolving capability is far more powerful than quadrupole-based instruments, there is a sacrifice in sensitivity at extremely high resolution, which can often translate into a degradation in detection capability for some elements, compared to other spectral interference correction approaches. A full review of magnetic sector technology for ICP-MS is given in Chapter 8.

MATRIX INTERFERENCES

Let us now take a look at the other class of interference in ICP-MS— suppression of the signal by the matrix itself. There are basically three types of matrix-induced interferences. The first and simplest to overcome is often called a sample transport effect and is a physical suppression of the analyte signal, brought on by the level of dissolved solids or acid concentration in the sample. It is caused by the sample's impact on droplet formation in the nebulizer or droplet size selection in the spray chamber. In the case of organic matrices, it is usually caused by variations in the pumping rate of solvents with different viscosities. The second type of matrix suppression is caused when the sample affects the ionization conditions of the plasma discharge. This results in the signal being suppressed by varying amounts, depending on the concentration of the matrix components. This type of interference is exemplified when different concentrations of acids are aspirated into a cool plasma. The ionization conditions in the plasma are so fragile that higher concentrations of acid result in severe suppression of the analyte signal. This can be seen very clearly in Figure 14.5, which shows sensitivity for a selected group of elements in varying concentrations of nitric acid in a cool plasma (12).

Compensation Using Internal Standardization

The classic way to compensate for a physical interference is to use internal standardization (IS). With this method of correction, a small group of elements (usually at the ppb level) are spiked into the samples, calibration stan-

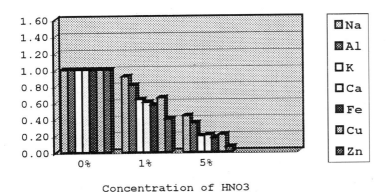

Concentration of HNO3

FIGURE 14.5 Matrix suppression caused by increasing concentrations of HNO$_3$, using cool plasma conditions (RF power: 800 W, nebulizer gas: 1.5 L/min). (From Ref. 12.)

dards, and blank to correct for any variations in the response of the elements caused by the matrix. As the intensity of the internal standards changes, the element responses are updated, every time a sample is analyzed. The following criteria are typically used for selecting the internal standards:

- They are not present in the sample.
- The sample matrix or analyte elements do not spectrally interfere with them.
- They do not spectrally interfere with the analyte masses.
- They should not be elements that are considered environmental contaminants.
- They are usually grouped with analyte elements of a similar mass range. For example, a low-mass internal standard is grouped with the low-mass analyte elements and so on up the mass range.
- They should be of a similar ionization potential to the groups of analyte elements so they behave in a similar manner in the plasma.
- Some of the most common elements/masses reported to be good candidates for internal standards include ^{9}Be, ^{45}Sc, ^{59}Co, ^{74}Ge ^{89}Y, ^{103}Rh, ^{115}In, ^{169}Tm, ^{175}Lu, ^{187}Re, and ^{232}Th.

A simplified representation of internal standardization is seen in Figure 14.6, which shows updating the analyte response curve across the full mass range, based on the intensities of low-, medium-, and high-mass internal standards. It should also be noted that internal standardization is also used to compensate for long-term, signal drift produced by matrix components

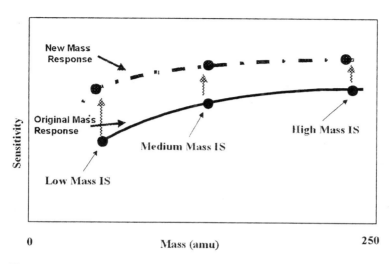

Sensitivity

New Mass
Response

Original Mass
Response

Low Mass IS

Medium Mass IS

High Mass IS

0   Mass (amu)   250

FIGURE 14.6   The analyte response curve is updated across the full mass range, based on the intensities of low-, medium-, and high-mass internal standards.

slowly blocking the sampler and skimmer cone orifices. Although total dissolved solids are usually kept below 0.2% in ICP-MS, this can still produce instability of the analyte signal over time with some sample matrices. It should also be emphasized that the difference in intensities of the internal standard elements across the mass range will indicate the flatness of the mass response curve. The flatter the mass response curve (i.e., less mass discrimination) the easier it is to compensate for matrix-based suppression effects using internal standardization.

Space-Charge-Induced Matrix Interferences

Many of the early researchers reported that the magnitude of signal suppression in ICP-MS increased with decreasing atomic mass of the analyte ion (13). More recently it has been suggested that the major cause of this kind of suppression is the result of poor transmission of ions through the ion optics due to matrix-induced space-charge effects (14). This has the effect of defocusing the ion beam, which leads to poor sensitivity and detection limits, especially when trace levels of low-mass elements are being determined in the presence of large concentrations of high-mass matrices. Unless any compensation is made, the high-mass matrix element will dominate the ion beam, pushing the lighter elements out of the way (15). This can be seen in Figure 14.7, which shows the classic space-charge effects of a uranium (major isotope

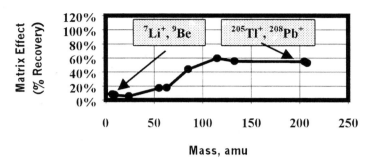

Mass, amu

FIGURE 14.7 Space-charge matrix suppression caused by 1000 ppm uranium is significantly higher on low-mass elements such as Li and Be than it is with the high-mass elements such as Tl and Pb. (From Ref. 15.)

$^{238}U^+$) matrix on the determination of $^7Li^+$, $^9Be^+$, $^{24}Mg^+$, $^{55}Mn^+$, $^{85}Rb^+$, $^{115}In^+$, $^{133}Cs^+$, $^{205}Tl^+$, and $^{208}Pb^+$. It can clearly be seen that the suppression of the low-mass elements such as Li and Be is significantly higher than with the high-mass elements such as Tl and Pb in the presence of 1000 ppm uranium.

There are a number of ways to compensate for space-charge matrix suppression in ICP-MS. Internal standardization has been used, but unfortunately it does not address the fundamental cause of the problem. The most common approach used to alleviate or at least reduce space-charge effects is to apply voltages to individual lens components of the ion optics. This is achieved in a number of different ways, but irrespective of the design of the ion focusing system, its main function is to reduce matrix-based suppression effects, by steering as many of the analyte ions through to the mass analyzer, while rejecting the maximum number of matrix ions. For more details on space-charge effects and different designs of ion optics, refer to Chapter 6 on the ion focusing system.

REFERENCES

1. Vaughan MA, Horlick G. Appl Spectrosc 1987; 41(4):523.
2. Tan SN, Horlick G. Appl Spectrosc 1986; 40(4):445.
3. Douglas DJ, French JB. Spectrochim Acta 1986; 41B(3):197.
4. Isotopic composition of the elements. Pure Appl Chem 1991; 63(7):991–1002 (IUPAC).
5. Willie SN, Iida Y, McLaren JW. At Spectrosc 1998; 19(3):67.
6. Jiang SJ, Houk RS, Stevens MA. Anal Chem 1988; 60:1217.

7. Sakata K, Kawabata K. Spectrochim Acta 1994; 49B:1027.
8. Tanner SD, Paul M, Beres SA, Denoyer ER. At Spectrosc 1995; 16(1):16.
9. Neubauer KR, Wolf RE. Determination of arsenic in chloride matrices. Perkin-Elmer Instrum Appl Note, 2000.
10. Hutton R, Walsh A, Milton D, Cantle J. ChemSA 1992; 17:213–215.
11. Tittes W, Jakubowski N, Stuewer D. Poster Presentation at Winter Conference on Plasma Spectrochemistry, San Diego, 1994.
12. Collard JM, Kawabata K, Kishi Y, Thomas R. Micro, January 2002.
13. Olivares JA, Houk RS. Anal Chem 1986; 58:20.
14. Tanner SD, Douglas DJ, French JB. Appl Spectrosc 1994; 48:1373.
15. Tanner SD. J Anal At Spectrom 1995; 10:905.

15

Contamination Issues

Serious consideration must be given to contamination issues in ICP-MS, particularly in the area of sample preparation. If you have been using flame AA or ICP-OES, you will probably have to rethink your sample-preparation procedures for ICP-MS. This chapter takes a closer look at the major causes of contamination and analyte loss in ICP-MS and how they affect both the analysis and the method development process.

There are many factors that influence the ability to get the correct result with any trace element technique. Unfortunately, with ICP-MS, the problem is magnified even more because of its extremely high sensitivity. To ensure that the data reported is an accurate reflection of the sample in its natural state, the analyst must be not only aware of all the potential sources of contamination, but also the many reasons why analyte loss is a problem in ICP-MS. Figure 15.1 shows the major factors that can impact the analytical result in ICP-MS.

COLLECTING THE SAMPLE

Collecting the sample and maintaining its integrity is a science all of its own and is beyond the scope of this book. However, it is worth discussing briefly to understand its importance in the overall scheme of collecting, preparing, and analyzing the sample. The object of sampling is to collect a portion of the material that is small enough to be conveniently transported and handled while still accurately representing the bulk material being sampled. Depending on the sampling requirements and the type of sample, there are basically three main types of sampling procedures. They are:

> Random sampling is the most basic type of sampling and represents only the composition of the bulk material at the time and place it was sampled. If the composition of the material is known to vary with

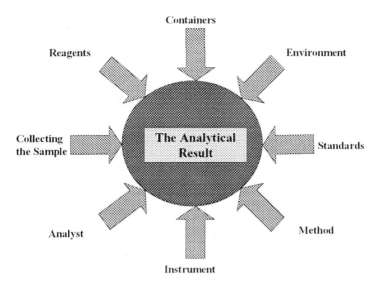

FIGURE 15.1   Major factors that can influence the analytical result in ICP-MS.

time, individual samples collected at suitable intervals and analyzed
separately can reflect the extent, frequency, and duration of these
variations

Composite sampling is when a number of samples are collected at the
same point, but at different times, and mixed together before being
analyzed

Integrated sampling is achieved by mixing together a number of
samples, which have been collected simultaneously from different
points

We will not go into which type of sampling is the most effective, but just
to emphasize that unless the correct sampling or subsampling procedure is
used, the analytical data generated by the ICP-MS instrumentation is
seriously flawed because it may not represent the original bulk material. If
the sample is a liquid, it is also important to collect the sample in clean
containers (see later), which have been thoroughly washed out beforehand. In
addition, if the sample is being kept for a long period of time before analysis, it
is essential that the analytes are kept in solution by using some kind of
preservative such as a dilute acid (this will also help to stop the analytes being
absorbed into the walls of the container). It is also important to keep the
samples as cool as possible to avoid evaporation losses. Kratochvil and

Taylor give an excellent review of the importance of sampling for chemical analysis (1).

PREPARING THE SAMPLE

As mentioned previously, ICP-MS was originally developed for the analysis of liquid samples. If the sample is not a liquid, some kind of sample preparation has to be carried out to get it into solution. There is no question that collecting a solid sample, preparing it, and getting it into solution probably represents the most crucial steps in the overall ICP-MS analytical methodology, because of the potential sources of contamination from grinding, sieving, weighing, dissolving, and diluting the sample. Let us take a look at these steps in greater detail and in particular focus on their importance when being used for ICP-MS.

GRINDING THE SAMPLE

Some fine powder solid samples are ready to be dissolved without grinding, but merely by passing them through a fine mesh sieve (mesh is typically 0.1–0.2 mm^2 mesh). Other types of coarser solid samples, such as soils, need to be first passed through a coarse mesh sieve (typically 2 mm^2 mesh) to be ready for dissolution (2). However, if the solid sample is not in a convenient form to be dissolved, it has to ground to a smaller particle size. The main reason for this is to improve the homogeneity of the original sample taken, as well as to make it more representative when taking a subsample. The ideal particle size will vary depending on the sample, but is typically ground to pass through a fine mesh sieve (0.1-mm^2 mesh). This uniform particle size ensures that the particles in the "test portion" are the same size as the particles in the rest of the ground sample. Another reason for grinding the sample into small uniform particles is that it makes it easier to dissolve.

   The process of grinding a sample with a pestle and mortar or a ball mill and passing it through a metallic sieve can be a huge source of contamination. This can originate from a previous sample that was being prepared or from materials used in the manufacture of the grinding or sieving equipment. For example, if tungsten carbide equipment is used to grind the sample, major elements such as tungsten and carbon as well as additive elements such as cobalt and titanium can also be a problem. Additionally, sieves, which are made from stainless steel, bronze, or nickel, can also introduce metallic contamination into the sample. To minimize some of these problems, plastic sieves are often used. However, it does not get around the problem of contamination from the grinding equipment. For this reason, it is usual to discard the first portion of the sample or even to use different grinding and sieving equipment for different kinds of samples.

SAMPLE DISSOLUTION METHODS

Unfortunately, there is not one dissolution procedure that can be used for all types of solid samples. There are many different approaches used to get solid samples into solution. For some samples, it is fairly straightforward and fast, while for others it can be very complex and time-consuming. However, all the successful sample dissolution procedures used in ICP-MS usually have a number of things in common:

> Complete dissolution is desired.
> Ultrapure reagents are used.
> The best reagents should not interfere with the analysis.
> There should be no loss of analyte.
> No chemical attack or corrosion of reaction or dilution containers.
> Safety is paramount.
> Ideally, it should be fast.

Even though the contamination issues are exaggerated with ICP-MS, the most common approaches used to get samples into solution are going to be very similar to the ones used for other trace element techniques. The most common dissolution techniques include:

> Hot plate, pressure bombs (3), or microwave digestion (4) using a concentrated acid/oxidizing agent such as nitric acid (HNO_3), perchloric acid ($HClO_4$), hydrofluoric acid, aqua regia, hydrogen peroxide, or various mixtures of them—these are among the most common approaches to dissolution and are typically used for metals, soils/sediments (5), minerals (6), and biological samples (7).
> Dissolution with strong bases such as caustic or trimethyl ammonium hydroxide (TMAH)—typically used for biological samples (8).
> Heating with fusion mixtures or fluxes such as lithium metaborate, sodium carbonate, or sodium peroxide in a metal crucible (e.g., platinum, silver, or nickel) and redissolving in a dilute mineral acid—typically used for ceramics, stubborn minerals, ores, rocks, and slags (9,10).
> Dry ashing using a flame, heat lamp, or a heated muffle furnace and redissolving the residue in a dilute mineral acid—typically used for organic or biological matrices (11).
> Wet ashing using concentrated acids (usually with some kind of heat)— typically used for organic/petrochemical/biomedical samples (12).
> Dissolution with organic solvents—typically used for organic/oil-type samples (13).

The choice of which one to use is often very complicated and depends on criteria such as the size of the sample, the matrix components in the sample, the elements to be analyzed, the concentration of the elements being determined, the types of interferences anticipated, the type of ICP-MS equipment being used, the time available for the analysis, safety concerns, and the expertise of the analyst. However, with ICP-MS the contamination issues are probably the greatest concern. For that reason, the most common approach to sample preparation is to keep it as simple as possible, because the more steps that are involved, the more chance there is of contaminating the sample. This means that ideally, if the sample is already a liquid, a simple acidification might be all that is needed. If the sample is a solid, a straightforward acid dissolution is preferred over the more complex and time-consuming fusion and ashing procedures. An excellent handbook of decomposition methods used for analytical chemistry was written by Bock in 1979 (14).

It is also important to emphasize that many acids that are used for AA and ICP-OES are not ideal for ICP-MS because of the polyatomic spectral interferences they produce. Although this is not strictly a contamination problem, it can significantly impact your data if not taken into consideration. For example, if vanadium or arsenic is being determined, it is advisable not to use hydrochloric acid (HCl) or $HClO_4$, because they generate the polyatomic ions such as $^{35}Cl^{16}O^+$ and $^{40}Ar^{35}Cl^+$, which interfere with the isotopes $^{51}V^+$ and $^{75}As^+$, respectively. Sulfuric acid (H_2SO_4) and phosphoric acid (H_3PO_4) are also acids that should be avoided if possible, because they generate sulfur-

TABLE 15.1 Typical Polyatomic Spectral Interferences Generated by Common Mineral Acids and Dissolution Chemicals

Acid/solvent/fusion mixture	Interference	Element/isotope
HCl	$^{35}Cl^{16}O^+$	$^{51}V^+$
HCl	$^{40}Ar^{35}Cl^+$	$^{75}As^+$
HNO_3	$^{14}N^{14}N^+$	$^{28}Si^+$
HNO_3	$^{14}N^{14}N^{16}O^+$	$^{44}Ca^+$
HNO_3	$^{40}Ar^{15}N^+$	$^{55}Mn^+$
H_2SO_4	$^{32}S^{16}O^+$	$^{48}Ti^+$
H_2SO_4	$^{34}S^{18}O^+$	$^{52}Cr^+$
H_2SO_4	$^{32}S^{16}O^{16}O^+$	$^{64}Zn^+$
H_3PO_4	$^{31}P^{16}O^{16}O^+$	$^{63}Cu^+$
Any organic solvent	$^{12}C^{12}C^+$	$^{24}Mg^+$
Any organic solvent	$^{40}Ar^{12}C^+$	$^{52}Cr^+$
Lithium-based fusion mixtures	$^{40}Ar^{7}Li^+$	$^{47}Ti^+$
Boron-based fusion mixtures	$^{40}Ar^{11}B^+$	$^{51}V^+$
Sodium-based fusion mixtures	$^{40}Ar^{23}Na^+$	$^{63}Cu^+$

and phosphorus-based polyatomic ions. For this reason, if there is a choice of which acid to use for dissolution, HNO_3 is the preferred one to use. Even though it can generate interferences of its own, they are generally less severe than those of the other acids (15). Table 15.1 shows the kinds of polyatomic spectral interferences generated by the most common mineral acids and dissolution chemicals.

In addition, fusion mixtures present unique problems for ICP-MS, not only because the major elements form polyatomic spectral interferences with the argon gas, but the elevated levels of dissolved solids in the sample can cause blockage of the interface cones, which over time can lead to signal drift. An additional problem with a fusion procedure is the risk of losing volatile analytes because of the high temperature of the muffle furnace or flame used to heat the open crucible.

CHOICE OF REAGENTS AND STANDARDS

Careful consideration must be given to the choice and purity of reagents, especially if sub-ppt concentration levels are expected. General laboratory- or reagent-grade chemicals used for AA or ICP-OES sample preparation are not usually pure enough. For that reason, most manufacturers of laboratory chemicals now offer ultra-high-purity grades of chemicals, acids, and fusion mixtures specifically for use with ICP-MS. It is therefore absolutely essential that the highest grade chemicals and water be used in the preparation and dilution of the sample. In fact, the grade of deionized water used for dilution and the cleaning of vessels and containers is very important in ICP-MS. Less pure water such as single-distilled or deionized water is fine for flame AA or ICP-OES, but is not suitable for use with ICP-MS because it could possibly contain contaminants such as dissolved inorganic/organic matter, suspended dust/scale particles, and possibly microorganisms. All these contaminants can affect reagent blank levels and negatively impact instrument and method detection limits. This necessitates using the most chemically pure water for ICP-MS work. There are several water-purification systems on the market that use combinations of filters, ion exchange cartridges, and/or reverse osmosis systems to remove the particulates, organic matter, and trace metal contaminants. These ultra-high-purity water systems (similar to the ones used for semiconductor processing) typically produce water with a resistance of better than 18 Megohms (16).

Another area of concern with regard to contamination is in the selection of calibration standards. Because ICP-MS is a technique capable of quantifying over 70 different elements, it will be detrimental to the analysis to use calibration standards that are developed for a single-element technique such as atomic absorption. These single-element standards are certified only for the

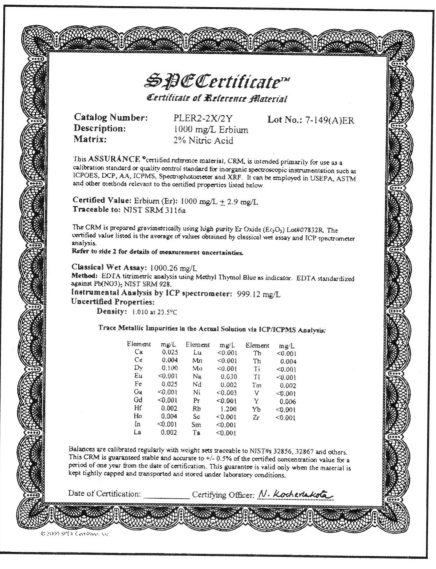

$\mathcal{SPECertificate}^{TM}$

Certificate of Reference Material

Catalog Number: PLER2-2X/2Y Lot No.: 7-149(A)ER
Description: 1000 mg/L Erbium
Matrix: 2% Nitric Acid

This ASSURANCE *certified reference material, CRM, is intended primarily for use as a calibration standard or quality control standard for inorganic spectroscopic instrumentation such as ICPOES, DCP, AA, ICPMS, Spectrophotometer and XRF. It can be employed in USEPA, ASTM and other methods relevant to the certified properties listed below.

Certified Value: Erbium (Er): 1000 mg/L ± 2.9 mg/L
Traceable to: NIST SRM 3116a

The CRM is prepared gravimetrically using high purity Er Oxide (Er$_2$O$_3$) Lot#07832R. The certified value listed is the average of values obtained by classical wet assay and ICP spectrometer analysis.
Refer to side 2 for details of measurement uncertainties.

Classical Wet Assay: 1000.26 mg/L
Method: EDTA titrimetric analysis using Methyl Thymol Blue as indicator. EDTA standardized against Pb(NO3)$_2$ NIST SRM 928.
Instrumental Analysis by ICP spectrometer: 999.12 mg/L
Uncertified Properties:
    Density: 1.010 at 23.5°C

Trace Metallic Impurities in the Actual Solution via ICP/ICPMS Analysis:

Element	mg/L	Element	mg/L	Element	mg/L
Ca	0.025	Lu	<0.001	Th	<0.001
Ce	0.004	Mn	<0.001	Th	0.004
Dy	0.100	Mo	<0.001	Ti	<0.001
Eu	<0.001	Na	0.030	Tl	<0.001
Fe	0.025	Nd	0.002	Tm	0.002
Ga	<0.001	Ni	<0.003	V	<0.001
Gd	<0.001	Pr	<0.001	Y	0.006
Hf	0.002	Rb	1.200	Yb	<0.001
Ho	0.004	Sc	<0.001	Zr	<0.001
In	<0.001	Sm	<0.001		
La	0.002	Ta	<0.001		

Balances are calibrated regularly with weight sets traceable to NIST#s 32856, 32867 and others. This CRM is guaranteed stable and accurate to +/- 0.5% of the certified concentration value for a period of one year from the date of certification. This guarantee is valid only when the material is kept tightly capped and transported and stored under laboratory conditions.

Date of Certification: _____ Certifying Officer: N. Kocherlakota

© 2000 SPEX CertiPrep, Inc.

FIGURE 15.2 Certificate for a 1000 mg/L erbium certified reference standard used in ICP-MS, showing values for over 30 trace metal contaminants (Courtesy of SPEX Certiprep.)

analyte element and not for any others. It is therefore absolutely critical to use calibration standards that have been specifically made for a multielement technique such as ICP-MS. It does not matter whether they are single- or multielement standards, as long as the certificate contains information on the suite of analyte elements you are interested in and any other potential interferents. It is also desirable if the certified values have been confirmed by both a classical wet technique and an instrumental technique, all of which are traceable to National Institute of Standards and Technology (NIST) reference material. It is also important to fully understand the uncertainty or error associated with a certified value, so you know how it impacts the data you report (17). Figure 15.2 is a certificate for a 1000 mg/L erbium certified reference standard used in ICP-MS, showing values for over 30 trace metal contaminants.

The same also applies if a calibration standard is being made from a high-purity salt of the metal. The salt has to be certified not only for the element of interest, but also for the full suite of analyte elements and also other elements that could be potential interferents. It is also important to understand the shelf life of these standards and chemicals, and how long-term storage impacts the concentration of the analyte elements, especially at such low levels.

VESSELS, CONTAINERS, AND SAMPLE-PREPARATION EQUIPMENT

The containers used for preparation, dilution, storage, and introduction of the sample can have a huge impact on your data in ICP-MS. Traditional glassware such as beakers, volumetric flasks, and autosampler tubes, which are fine for AA and ICP-OES work, are not ideally suited for ICP-MS. The major problem is potential contamination from the major elemental components of the glassware. For example, glass made from soda lime contains percent concentrations of silicon, sodium, calcium, magnesium, and aluminum, while borosilicate glass contains high levels of boron. Besides these major elements, they might also contain minor concentrations of Zr, Li, Ba, Fe, K, and Mn. Unfortunately, if the sample solution is highly acidic, there is a strong possibility that these elements can be leached out of the glassware. In addition to the contamination issues, analytes can be absorbed into the walls of volumetric flasks and beakers made of glass. This can be a serious problem if the sample or standard is being stored for extended periods of time, especially if the analyte concentrations are extremely low. If using glassware is unavoidable, it is a good idea to clean the glassware on a regular basis using chromic acid and/or some kind of commercial glass detergent such as Decon™ or Citranox™. If long-term storage is a necessity, either avoid using glassware or minimize the analyte loss by keeping the solutions acidified (~pH 2), so there is very little chance of absorption into the walls of the glass (18).

Glassware is such a universal material used for sample preparation that it is very difficult to completely avoid it. However, serious consideration should be given to looking for alternative materials in as many of the ICP-MS sample-preparation steps as possible. Today, the most common materials used to manufacture beakers, volumetric containers, and autosampler tubes for ultra trace element techniques such as GFAA and ICP-MS are mainly plastic based. Over the past 10–15 years, the demand for these kinds of materials has increased significantly because of the contamination issues associated with glassware. Some plastics are more inert and more pure than others, so thought should be given to which one is optimal for your samples. Selection should be made based on the suite of elements being analyzed, analyte concentration levels, matrix components, or whether it is an aqueous-, acid- or organic-based solution. Some of the most common plastic materials used in the manufacture of sample-preparation vessels and/or sample-introduction components include polypropylene (PP), polyethylene (PE), polysulfide (PS), polycarbonate (PC), polyvinyl chloride (PVC), polyvinylfluoride (PVF), PFA (perfluoroalkoxy), and polytetrafluoroethylene (PTFE). It is generally felt that PTFE or PFA probably represent the cleanest materials, and even though they are the most expensive, they are considered the most suitable for ultra trace ICP-MS work. However, even though these types of plastics are generally much cleaner than glass, they still contain some trace elements. For example, certain plastics might contain phosphorus from the mold-releasing agent, or some plastic tube caps and covers are manufactured with barium compounds to enhance their color. These are all potential sources of contamination, which could cause serious problems in ICP-MS, especially if heat is involved in sample preparation. This is particularly true if microwave dissolution is used to prepare the sample, because of the potential for high-temperature breakdown of the polymer material over time. Table 15.2, which was taken from a publication about 20 years ago, gives trace element contamination levels of some common plastics used in the manufacture of laboratory beakers, volumetric ware, and autosampler tubes (19). It should be strongly emphasized that these data might not be representative of today's products, but should be used only as an approximation for comparison purposes.

Even though microwave dissolution is rapidly becoming the sample-dissolution method of choice over conventional hot plate digestion methods, it will not be discussed in great detail in this chapter. Such is the maturity and proven capability of this approach nowadays that there are a multitude of textbooks and reference papers in the public domain covering a wide range of

TABLE 15.2 Typical Trace Element Contamination Levels of Some Common Plastic Materials Used in the Manufacture of Laboratory Beakers, Volumetric Ware, and Autosampler Tubes

Material	Na (ppm)	Al (ppm)	K (ppm)	Sb (ppm)	Zn (ppm)
Polyethylene (CPE)	1.3	0.5	5	0.005	—
Polyethylene (LPE)	15	30	0.6	0.2	520
Polypropylene (PP)	4.8	55	—	0.6	—
Polysulfide (PS)	2.2	0.5	—	—	—
Polycarbonate (PC)	2.7	3.0	—	—	—
Polyvinylchloride (PVC)	20	—	—	—	—
Polytetrafluoroethylene (PTFE)	0.16	0.23	90	—	—

Source: Ref. 19.

samples being analyzed by ICP-MS, including geological materials (20), soils (21), sediments, (22), waters (23), biological materials (24), and foodstuffs (25).

However, irrespective of which digestion method is used, consideration should also be given to other equipment and materials used in sample preparation, because the decision on what to use can impact the analysis. Some of these potential areas of concern include:

> The quality of the filtering materials if the particulates need to be filtered out. For example, should conventional filter papers be used or ones made from cellulose or acetate glass, or should vacuum filtration using sintered discs be used instead?
>
> If blood is being drawn for analysis, the cleanliness of the syringe and the material it is made from can contribute to contamination of the sample.
>
> Paper towels are used for many different reasons in a laboratory. These are generally high in zinc and contain trace levels of transition metals such as Fe, Cr, and Co, so avoid using them in and around your sample prep areas.
>
> Pipettes, pipette tips, and suction bulbs can all contribute to trace metal contamination levels, so for that reason the disposable variety is recommended.

It is important to emphasize that whatever containers, vessels, beakers, volumetric ware, or equipment is used to prepare the sample for ICP-MS analysis, it is absolutely critical that when not in use, they are soaked and washed in a dilute acid (1–2% HNO_3 is typical). In addition, if they are not being used for extended periods, they should be stored with dilute acid in them. Wherever possible, disposable equipment such as autosampler tubes and pipette tips should be used and thrown away after use to cut down on contamination.

THE ENVIRONMENT

The environment in this case refers to the cleanliness of the surrounding area where the instrumentation is installed, where sample preparation is carried out, or any other area the sample comes in contact with. It is advisable that the sample-preparation area is as close to the instrument as possible, without actually being in the same room, so that the sample is not exposed to any additional sources of contamination. It is recommended that dissolution is carried out in clean, metal-free fume extraction hoods and if possible in a separate area to samples that are being prepared for less sensitive techniques such as flame AA or ICP-OES. In addition to having a clean area for dissolution, it is also important to carry out other sample-preparation tasks such as weighing, filtering, pipetting, diluting, etc. in a clean environment.

These kinds of environmental contamination problems are an everyday occurrence for the semiconductor industry because of the strict cleanliness demands required for the fabrication of silicon wafer and production of semiconductor devices. The purity of the silicon wafers has a direct effect on the yield of devices, so it is crucial that trace element contamination levels are kept to a minimum to reduce defects. This means that any analytical methodology used to determine purity levels on the surface of silicon wafers, or in the high-purity chemicals used to manufacture the devices, must be spotlessly clean. These unique demands of the semiconductor industry has led to the development of special air-filtration systems, which continually pump the air through ultraclean HEPA filters to remove the majority of airborne particulates.

The efficiency of particulate removal will depend on the analytical requirements, but for the semiconductor industry, it is typical to work in environments that contain 1 or 10 particles (<0.2 µm) per cubic foot of air (Class 1 and 10 clean rooms, respectively). These kinds of precautions are absolutely necessary to maintain low instrument background levels for the analysis of semiconductor-related samples, but might not be required for other types of applications. So even though contamination-free analysis is important, it might be suffice to work in a Class 100, 1000, or 10000 clean room and still meet your cleanliness objectives (26).

These clean rooms tend to be very expensive to build, so if your budget does not stretch to a "full-blown" clean room, it might be worth investing in special HEPA filter enclosures just for your instrument and sample-preparation area. These are typically either mobile units that can be wheeled around the laboratory and placed around different equipment or hood-based enclosures that are placed over a particular instrument. Whatever system is used, their objective is to ensure that the area around the equipment is free of airborne contamination and instrument background levels are as low as possible.

THE ANALYST

The expertise of the analyst who actually prepares the samples and carries out the analysis can be a major factor in getting the right result by ICP-MS. Even if all precautions have been taken to cut down on contamination, if the analyst is not experienced in working with ICP-MS and does not understand all the potential pitfalls, the analysis could be doomed to failure. For example, they have to be aware of all the potential contaminants that are generated by their own bodies or the clothes or jewelry they are wearing. Table 15.3 shows some common trace elements found on the human body. It is by no means an exhaustive list, but at least it gives you an idea of the problem.

These kinds of personal contamination problems are the reason you often see operators of equipment used in the semiconductor industry wearing "bunny suits." These are white suits that cover the entire body of the operator, including head, hands, and feet, to stop any human-based contamination getting into the equipment or instrumentation. They are not so important for higher levels of quantitation, but are absolutely necessary for the kind of ultra trace contamination levels found in the electronics industry.

INSTRUMENT AND METHODOLOGY

The instrument and the methodology itself can also be a potential source of error. It is therefore important to be aware of this and to understand what is required when developing a method to carry out the determination of ultra

TABLE 15.3 Some Common Trace Elements Contaminants Found on and Around the Human Body

Source of contamination	Trace metal contaminant
Hair	Zn, Cu, Fe, Pb, Mn
Skin	Zn, Cu
Nails	Ca, Si
Jewelry	Au, Ag, Cu, Fe, Ni, Cr
Cigarette Smoke	Cd, As, K, Fe, B
Cosmetics	Zn, Bi
Deodorants	Al

trace levels by ICP-MS. As mentioned previously, the choice of sample-preparation methodology can impact the analysis by either causing corrosion problems for some of the instrument components, producing spectral interferences on the analyte, or creating matrix-induced signal drift problems. However, in addition to optimizing sample preparation, a great deal of thought must also go into the choice of instrumental components and to understand how they impact the method development process. Some of the criteria that should be under consideration when deciding on the analytical methodology include the following:

The acid concentration in the final solution being presented to the instrument ideally should be 2–3% maximum because of the sample transport interferences associated with high concentrations of mineral acids.

If highly corrosive acids such as hydrofluoric acid are being used, the appropriate corrosion-resistant sample-introduction components should be used such as a plastic spray chamber and nebulizer, sapphire sample injector, and platinum interface cones.

Hydrochloric, sulfuric, and phosphoric acids should be avoided because of the spectral problems created by the high concentration of chlorine, sulfur, and phosphorus ions in the matrix.

The choice of fusion mixture should be given serious consideration because of the potential for the lithium-, sodium-, or potassium-based salts to deposit themselves around the sampler or skimmer cone orifice, which over time can lead to serious drift problems.

The sample weight might have to be compromised if a fusion mixture is required, because 0.2% is the maximum level of dissolved solids that can be aspirated into the ICP mass spectrometer.

There are many grades of argon gas available for spectrochemical analysis. For ultra trace determinations by ICP-MS, the highest grade should always be used (usually ultra high purity grade argon is 99.99999% pure).

Petrochemical-type samples usually require the addition of oxygen to the nebulizer gas flow to "burn-off" the organic matrix, so the highest quality of the oxygen should be used.

The choice of pneumatic tubing should be compatible with the sample solution. For example, when analyzing organic samples, suitable pump tubing and sample capillary should be used that is resistant to the organic solvent.

There are many different kinds of pump tubing and capillary. If a polyvinyl-chloride-based tubing is being used, chlorine could potentially be leached out and cause spectral interferences.

Peristaltic pump speed, washout times, read delays, and stability times should be optimized based on the sample matrix and suite of elements because of memory effects in the sample introduction/interface areas and therefore the possibility of contamination from the previous sample being analyzed

What are the expected analyte concentrations and matrix levels? This will impact whether the sample can be diluted or whether the analytes need to be preconcentrated or the matrix components removed?

If the samples are completely unknown, it is a good strategy to dilute the sample 1:100 and get an approximation of the analyte concentrations using the instrument's "semiquant" routine. This can also give you an insight into understanding the potential interferences from the other elements in the sample

These are generally considered some of the most important criteria when deciding on an analytical methodology to analyze a set of samples by conventional solution nebulization. However, it should be emphasized that the strategy might also include the use of sampling accessories, such as laser ablation or flow injection. For example, the ability to analyze a solid directly by laser ablation eliminates most of the contamination issues with the preparation, dilution, and aspiration of liquid samples. Even though this might sound attractive, solid sampling has unique problems of its own. So before this approach is chosen, it is important to also understand all its limitations, especially for a particular set of samples. On the other hand, if solution nebulization is the preferred approach, will there be any benefit of using segmented flow analysis to reduce the amount of matrix entering the mass spectrometer? Clearly, for some matrices it is advantageous, while for others it might not be worth the effort. It is therefore important to understand these issues before a decision is made (refer to Chapter 17 on sampling accessories for more information).

Whatever analytical methodology approach is used, the issue of contamination must always be at the forefront of the decision. ICP-MS is such a sensitive technique that to take advantage of its unparalleled detection capability and sample throughput capabilities, analytical cleanliness, and optimized method development is of the utmost importance. If attention is paid to these areas, there is no question that data of the highest quality can be obtained, even at the ultra trace level. This chapter is not intended to be an exhaustive look at contamination or analyte loss issues, but just to make the reader aware that to get the right result in ICP-MS, it is important to examine all aspects of the analysis from first collection of the sample, all the way through to the quantitation by the instrument. If you are interested in finding out more about this subject, Ref. 27 is an excellent book on contamination control in trace metal analysis.

FURTHER READING

1. Kratochvil B, Taylor JK. Anal Chem 1981; 53(8):925A–938A.
2. EPA ICP-MS Method 200.8 for drinking water, ground water, waste water, sludges and soils, 1995, Environmental Protection Agency, Washington, DC.
3. Bernas B. Anal Chem 1986; 40(11):1586–1682.
4. Kingston HM, Jassie LB, eds. Introduction to Microwave Sample Preparation—Theory and Practice. American Chemical Society, 1988.
5. Hewitt A, Reynolds CM. At Spectrosc 1990; 11(5):187–192.
6. Nadkarni RA. Anal Chem 1984; 56:2233–2237.
7. Abu-Samra A. Anal Chem 1975; 47(8):1475–11477.
8. Pruszkowski E, Neubauer K, Thomas R. At Spectrosc 1998; 19(4):111–115.
9. Ingamells CO. Anal Chim Acta 1970; 52:323–334.
10. Belcher CB. Talanta 1963; 10:75–81.
11. Friend MT, Atomic Absorption Newsletter 1977; 16(2):46–49.
12. Bajo S, Suter U. Anal Chem 1982; 54(1):49–51.
13. McElroy F, Mennito A, Debrah E, Thomas R. Spectroscopy 1998; 13(2):42–53.
14. Bock R. A Handbook for Decomposition Methods in Analytical Chemistry. London, England: International Textbook Company Ltd., 1979.
15. Tan S, Horlick G. Appl Spectrosc 1986; 40:445.
16. Standard guide for ultrapure water used in the semiconductor and electronics industry, ASTM D-5127-98. American Society for testing and Materials, West Conshohoken, PA, 1998.
17. Kocherlakota N, Obernauf R, Thomas R. Spectroscopy 2002; 17(7).
18. Robertson DE. Anal Chem 1968; 40(7):1067–1072.
19. Moody JR, Lindstrom RM. Anal Chem 1977; 49(14):2264–2267.
20. Totland M, Jarvis I, Jarvis KE. Chem Geol 1992; 95:35–62.
21. Verma VL, McKee TM. Paper presented at the 7th Annual Waste Testing and Quality Assurance Symposium (EnvirACS), Washington, DC: Environmental Protection Agency, July 10, 1991.
22. ASTM Method No. D5258-92, Standard practice for acid extraction of elements from sediments using closed vessel microwave heating. Annual Book of ASTM Standards. West Conshohoken, PA: American Society for Testing and Materials, 1992. Note: A page number is not needed with an ASTM Method number.
23. ASTM Method No. D4309-91, Standard practice for sample digestion using closed vessel microwave heating technique for the determination of total recoverable metals in water. Annual Book of ASTM Standards. West Conshohoken, PA: American Society for Testing and Materials, 1991.
24. McCarthy HT, Ellis PC. J Anal Chem 1991; 74(3):566–569.
25. Sears D Jr, Grosser Z. Food Test Anal, June/July 1997.
26. Talasek T. Solid State Technol Dec 1993; 44–46.
27. Zief M, Mitchel JW. Contamination Control in Trace Metal Analysis. New York: John Wiley and Sons, 1976.

16

Routine Maintenance Issues

The components of an ICP-MS are generally more complex than other atomic spectroscopic techniques and as a result more time is required to carry out routine maintenance to ensure that the instrument is performing to the best of its ability. Some tasks involve a simple visual inspection of a part, while others involve cleaning or changing components on a regular basis. However, routine maintenance is such a critical part of owning an ICP-MS system that it can impact both the performance and the lifetime of the instrument.

The fundamental principle of ICP-MS is based on interfacing a plasma discharge at 10,000 K to a mass spectrometer at approximately 10^{-6} Torr. The sample is introduced in the form of a liquid aerosol (or solid particles with laser sampling) and then ionized in the plasma where the matrix and analyte ions are directed into the mass analyzer where they are separated and finally measured by the ion detection system. This principle, which gives ICP-MS its unequalled isotopic selectivity and sensitivity, unfortunately contributes to some of its weaknesses—the fact that the sample "flows into" the spectrometer and not "passed it" as with flame AA and ICP-OES. This means that the potential for thermal problems, corrosion, chemical attack, blockage, matrix deposits, and drift is much higher than with the other AS techniques. However, being fully aware of this fact and carrying out regular inspection of instrumental components can reduce and sometimes eliminate many of these potential problem areas. There is no question that a laboratory that initiates a routine preventative maintenance plan stands a much better chance of having an instrument that is ready and available for analysis whenever it is needed, compared to a laboratory that basically ignores these issues and assumes the instrument will look after itself.

Let us now look at the areas of the instrument that an owner needs to pay attention to. I will not go into great detail but just give a brief overview of what is important, so you can compare it with maintenance procedures of trace element techniques you are more familiar with. These areas should be very similar with all commercial ICP-MS systems, but depending on the design of the instrument and the types of samples being analyzed the regularity of changing or cleaning components might be slightly different (particularly if the instrument is being used for laser ablation work). The main areas that require inspection and maintenance on a routine or semiroutine basis include:

- Sample introduction system
- Plasma torch
- Interface region
- Ion optics
- Roughing pumps
- Air/water filters

Other areas of the instrument require less attention, but nevertheless the user should be aware of maintenance procedures required to extend their lifetime. They will be discussed at the end of this section.

SAMPLE INTRODUCTION SYSTEM

The sample introduction system, composed of the peristaltic pump, nebulizer, spray chamber, and drain system, takes the initial abuse from the sample matrix and, as a result, is an area of the ICP-MS that needs a great deal of care and attention. The principles of sample introduction area have been described in great detail in Chapter 3, so let us now examine what kind of maintenance it requires.

Peristaltic Pump Tubing

In ICP-MS, the sample is pumped at about 1 mL/min into the nebulizer, via a peristaltic pump. The constant motion and pressure of the pump rollers on the pump tubing, which is typically made from a polymer-based material, ensures a continuous flow of liquid to the nebulizer. However, over time, this constant pressure of the rollers on the pump tubing has the tendency to stretch it, which changes its internal diameter and therefore the amount of sample being delivered to the nebulizer. The impact is an erratic change in analyte intensity and a degrading of short-term stability.

As a result of this, the condition of the pump tubing should be examined every couple of days, particularly if your lab has a high sample workload or if extremely corrosive solutions are being analyzed. The peristaltic pump tubing is probably one of the most neglected areas, so it is absolutely essential that it be a part of your routine maintenance schedule. Here are some suggested tips to reduce pump tubing-based problems.

- Manually stretch new tubing before use.
- Maintain the proper tension on tubing.
- Ensure tubing is placed correctly in channel of the peristaltic pump.
- Periodically check flow of sample delivery—throw away tubing if in doubt.
- Replace tubing if there is any sign of wear—do not wait until it breaks.
- With high sample workload, change tubing every day or every other day.
- Release pressure on pump tubing when instrument is not in use.
- Pump and capillary tubing can be a source of contamination.
- Pump tubing is a consumable—keep a large supply of it on hand.

Nebulizers

The frequency of nebulizer maintenance will primarily depend on the types of samples being analyzed and the design of nebulizer being used. For example, in a crossflow nebulizer, the argon gas is directed at right angles to the sample capillary tip, in contrast to the concentric, where the gas flow is parallel to the capillary. This can be seen in Figures 16.1 and 16.2, which show schematics of a concentric and crossflow nebulizer, respectively.

The larger diameter of the liquid capillary and longer distance between the liquid and gas tips of the crossflow design make it far more tolerant to dissolved solids and suspended particles in your sample than the concentric

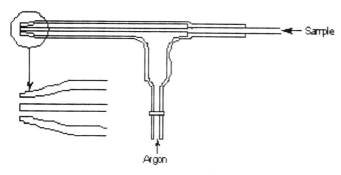

FIGURE 16.1 Schematic of a concentric nebulizer. (Courtesy of Meinhard Glass Products.)

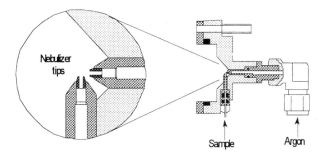

FIGURE 16.2 Schematic of a crossflow nebulizer. (Courtesy of PerkinElmer Life and Analytical Sciences.)

design. On the other hand, aerosol generation of a crossflow nebulizer is far less efficient than a concentric nebulizer and therefore produces less optimum size droplets required for the ionization process. As a result, concentric nebulizers generally produce higher sensitivity and slightly better precision than the crossflow design but are more prone to clogging.

So the choice of which nebulizer to use is usually based on the types of samples being aspirated and the data quality objectives of the analysis. However, whichever one is being used, attention should be paid to the tip of the nebulizer to ensure it is not getting blocked. Sometimes microscopic particles can build-up on the tip of the nebulizer, without the operator noticing, which, over time, can cause a loss of sensitivity, imprecision, and poor long-term stability. In addition, O-rings and sample capillary can be affected by the corrosive solutions being aspirated, which can also degrade performance. For these reasons, the nebulizer should always be a part of the regular maintenance schedule. Some of the most common things to check include:

- Visually check nebulizer aerosol by aspirating water—a blocked nebulizer will usually result in an erratic spray pattern with lots of large droplets.
- Remove blockage by either using backpressure from argon line or dissolving the material by immersing nebulizer in an appropriate acid or solvent—an ultrasonic bath can sometimes be used to aid dissolution, but check with manufacturer first, in case it is not recommended. (Note: never stick any wires down the end of the nebulizer, because it could do permanent damage.)
- Ensure nebulizer is securely seated in spray chamber end cap.
- Check all O-rings for damage or wear.

- Ensure sample capillary is inserted correctly into sample line of nebulizer.
- Nebulizer should be inspected every 1–2 weeks, depending on workload.

Spray Chamber

By far the most common design of spray chamber used in commercial ICP-MS instrumentation is the double-pass design, which selects the small droplets by directing the aerosol into a central tube. The larger droplets emerge from the tube and, by gravity, exit the spray chamber via a drain tube. The liquid in the drain tube is kept at positive pressure (usually by way of a loop), which forces the small droplets back between the outer wall and the central tube and emerges from the spray chamber into the sample injector of the plasma torch. Scott double-pass spray chambers come in a variety of shapes, sizes, and materials but are generally considered the most rugged design for routine use. Figure 16.3 shows a double-pass spray chamber (made of a polymer material), coupled to a crossflow nebulizer.

The most important maintenance with regard to the spray chamber is to make sure that the drain is functioning properly. A malfunctioning or leaking drain can produce a change in the spray chamber backpressure, producing

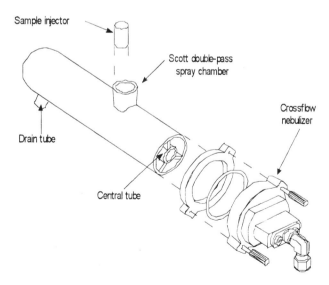

FIGURE 16.3 A double-pass spray chamber coupled to a crossflow nebulizer. (Courtesy of PerkinElmer Life and Analytical Sciences.)

fluctuations in the analyte signal, resulting in erratic and imprecise data. Less frequent problems can result from degradation of O-rings between the spray chamber and the sample injector of the plasma torch. Typical maintenance procedures regarding the spray chamber include.

- Make sure the drain tube fits tightly and that there are no leaks.
- Ensure the waste solution is being pumped from the spray chamber into the drain properly.
- If a drain loop is being used, make sure the level of liquid in the drain tube is constant.
- Check O-ring/ball joint between spray chamber exit tube and torch sample injector—make sure connection is snug.
- Spray chamber can be a source of contamination with some matrices/analytes, so flush thoroughly between samples.
- Empty spray chamber of liquid when instrument is not in use.
- Spray chamber and drain should be inspected every 1–2 weeks, depending on workload.

Plasma Torch

Not only is the plasma torch and sample injector exposed to the sample matrix and solvent, but it also has to sustain the analytical plasma at approx. 10,000 K. This combination makes for a very hostile environment and therefore is an area of the system that requires regular inspection and maintenance. As a result, one of the main problems is staining and discoloration of the outer tube of the quartz torch, due to heat and the corrosiveness nature of the liquid sample. If the problem is serious enough, it has the potential to cause electrical arcing. Another potential problem area is blockage of the sample injector from matrix components in the sample. As the aerosol exits the sample injector, desolvation takes place, which means that sample changes from small liquid droplets to minute solid particles prior to entering the base of the plasma. This is conceptually shown in Figure 16.4. Unfortunately with some sample matrices, these particles can deposit themselves on the tip of the sample injector over time, leading to possible clogging and drift. In fact, this can be a potentially serious problem when aspirating organic solvents, because carbon deposits can rapidly build up on the sample injector and cones unless a small addition of oxygen is made to the nebulizer gas flow.

Some torches use metal plates or shields to reduce the secondary discharge between the plasma and the interface. These are consumable items, because of the intense heat and the effect of the RF field on the shield. A shield in poor condition can affect instrument performance, so the user should always be aware of this and replace when necessary.

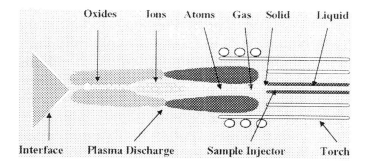

FIGURE 16.4   Rapid desolvation of the aerosol can lead to deposits on the tip of the sample injector.

Here are some useful maintenance tips with regard to the torch area:

- Look for discoloration or deposits on outer tube of quartz torch— remove material by soaking torch in appropriate acid or solvent if required.
- Check torch for thermal deformation of torch—a nonconcentric torch can cause loss of signal.
- Check sample injector for blockages—if the injector is demountable, remove material by immersing it in an appropriate acid or solvent if required (if the torch is one-piece, soak the whole torch in the acid).
- Ensure torch is positioned in the center of load coil and the correct distance from interface cone when replacing torch assembly.
- If the coil has been removed for any reason, make sure the gap between the turns is correct as per recommendations in operator's manual.
- Inspect any O-rings or ball joints for wear or corrosion—replace if necessary.
- If a shield or plate is used to ground the coil, ensure it is always in good condition—otherwise replace when necessary.
- Torch should be inspected every 1–2 weeks, depending on workload.

INTERFACE REGION

As the name suggests, the interface is the region of the ICP-MS, where the plasma discharge at atmospheric pressure is "coupled" to the mass spectrometer at 10^{-6} Torr by way of two interface cones—a sampler and skimmer. This coupling of such a high temperature ionization source such as an ICP to the metallic interface of the mass spectrometer puts unique demands on this region of the instrument, which is not experienced by any other AS technique. When this is combined with matrix, solvent, and analytes ions together with particulates and neutral species being directed at high velocity at the interface cones, it makes for an extremely harsh environment. The most common types of problems associated with the interface are blocking and/or corrosion of the sampler cone and, to a lesser extent, the skimmer cone. This is not always obvious, because often the build-up of material on the cone or corrosion around the orifice can take a long time to reveal itself. For that reason, the sampler and skimmer interface cones have to be inspected and cleaned on a regular basis. The frequency will often depend on the types of samples being analyzed and also the design of the ICP-MS spectrometer. For example, it is well documented that a secondary discharge at the interface can prematurely discolor and degrade the sampler cone, especially when complex matrices are being analyzed or if the instrument is being used for high sample throughput. The layout of a typical ICP-MS interface, showing the potential areas of blockage, is shown in Figure 16.5.

Besides the cones, the metal interface housing itself is also exposed to the high temperature plasma. For this reason, it needs to be cooled by a recirculating water system, usually containing some kind of antifreeze/corrosion inhibitor or by a continuous supply of mains water. Recirculating systems are probably more widely used, because the temperature at the interface can be controlled much better. There is no real routine maintenance involved with the interface housing, except maybe to check the quality of the coolant from time to time, to make sure there is no corrosion of the interface cooling system. If for any reason, the interface gets too hot, there are usually

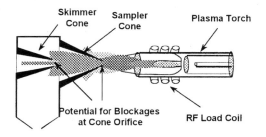

FIGURE 16.5 Layout of an ICP-MS interface showing potential areas of blockage. (Courtesy of Varian, Inc.)

built-in safety interlocks that will turn the plasma off. Some useful hints to prolong the lifetime of the interface and cones include:

- Check that both sampler and skimmer cone are clean and free of sample deposits—typical frequency is weekly, but it will depend on sample type and workload.
- If necessary remove and clean cones using manufacturers recommendations—typical approaches include immersion in a beaker of weak acid, or detergent placed in a hot water or ultrasonic bath. Abrasion with fine wire wool or a coarse polishing compound has also been used.
- Never stick any wire into the orifice—it could do permanent damage.
- Nickel cones will degrade rapidly with harsh sample matrices—use platinum cones for highly corrosive solutions and organic solvents.
- Periodically check cone orifice diameter and shape with a magnifying glass (10–20× magnification)—irregular shaped orifice will affect instrument performance.
- Thoroughly dry cones before installing them back in the instrument, because water/solvent could be pulled back into the mass spectrometer.
- Check coolant in recirculating system for signs of interface corrosion—such as copper or aluminum salts (or predominant metal of interface).

ION OPTICS

The ion optic system is usually positioned just behind or close to the skimmer cone to take advantage of the maximum number of ions entering the mass spectrometer. There are many different commercial designs and layouts, but they all have one thing in common and that is to transport the maximum number of analyte ions, while allowing the minimum number of matrix ions through to the mass analyzer. Figure 16.6 shows a typical layout of a traditional ion focusing system.

The ion focusing system is not traditionally thought of as a component that needs frequent inspection, but because of its proximity to the interface region, it can accumulate minute particulates and neutral species that, over time, can dislodge, find their way into the mass analyzer, and affect instrument performance. A dirty or contaminated ion optic system typically shows poor stability and/or a need to gradually increase lens voltages over time. For that reason, no matter what design of ion optics is used, inspection and cleaning every 2–3 months (depending on workload and sample type) should be an

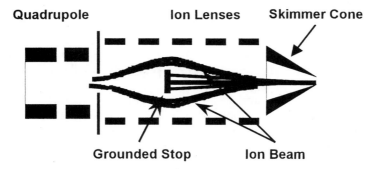

FIGURE 16.6   Layout of a traditional ion focusing system.

integral part of a preventative maintenance plan. Some useful tips for the ion optics in order to maintain maximum ion transmission and good stability include:

- Look for sensitivity loss over time, especially in complex matrices.
- If sensitivity is still low after cleaning sample introduction system, torch and interface cones, it could indicate that ion lens system is getting dirty.
- Try retuning or reoptimizing the lens voltages.
- If voltages are significantly different (usually higher than previous settings), it probably means lens components are getting dirty.
- When the lens voltages become unacceptably high, the ion lens system will probably need replacing or cleaning—use recommended procedures outlined in the operator's manual.
- Depending on the design of the ion optics, some single-lens systems are considered consumables and are discarded after a period of time. While multicomponent lens systems are usually cleaned using abrasive papers and/or polishing compounds and rinsed with water and an organic solvent.
- If cleaning ion optics, make sure they are thoroughly dry as water or solvent could be sucked back into the mass spectrometer.
- Gloves are usually recommended when reinstalling ion optic system, because of the possibility of contamination.
- Do not forget to inspect or replace O-rings or seals when replacing ion optics.
- Depending on instrument workload, you should expect to see some deterioration in the performance of the ion lens system after 3–4 months of use—this is a good approximation of when it should be inspected and cleaned or replaced if necessary.

ROUGHING PUMPS

There are typically two roughing pumps used in commercial instruments. One is used on the interface region and another used as a back-up to the turbo molecular pumps on the main vacuum chamber. They are usually oil-based rotary or diffusion pumps, which need new oil on a regular basis, depending on the instrument usage. The oil in the interface pump will need changing more often than the one on the main vacuum chamber, because it is pumping for a longer period of time. A good indication of when the oil needs to be changed is to look at its color in the "viewing glass." If it appears dark brown, there is a good chance that heat has broken down its lubricating properties and it needs to be changed. With the roughing pump on the interface, the oil should be changed every 1–2 months and with the main vacuum chamber pump, it should be changed every 3–6 months. These times are only approximations and will vary depending on the sample workload and the time the instrument is actually running. Some important tips when changing the roughing pump oil:

- Do not forget to turn the instrument and the vacuum off—if the oil is being changed from "cold," it might be useful to run the instrument for 10–15 min beforehand, to get the oil to flow better.
- Drain the oil into a suitable vessel—caution, the oil might be very hot if the instrument has been running all day.
- Fill the oil to the required level in the "viewing glass."
- Check for any loose hose connections.
- Replace oil filter if necessary.
- Turn instrument back on—check for any oil leaks around filling cap—tighten if necessary.

AIR FILTERS

Most of the electronic components, especially the ones in the RF generator are air-cooled. For this reason, the air filters should be checked, cleaned, or replaced on a fairly regular basis. Although this is not carried out as routinely as the sample introduction system, a typical time frame to inspect the air filters is every 3–6 months, depending on the workload and instrument usage.

OTHER COMPONENTS TO BE PERIODICALLY CHECKED

It is also important to emphasize that other components of the ICP-MS have a finite lifetime, which will need to be replaced or at least inspected from time to time. These components are not considered a part of the routine maintenance schedule and usually require a service engineer (or at least an experienced user) to clean or to change them. These areas include the following.

The Detector

Depending on usage and the levels of ion signals measured on a routine basis, the electron multiplier should last about 12 months. A failing detector will show itself as a rapid decrease in the "gain" setting, despite attempts to increase the detector voltage. The lifetime of a detector can be increased by avoiding measurements at masses that produce extremely high ion signals, such as those associated with the argon gas, solvent, or acid used to dissolve the sample (e.g., hydrogen, oxygen, and nitrogen) or any mass associated with the matrix itself. It is important to emphasize that when the detector is being replaced, it should be carried out by an experienced person wearing gloves, to reduce the possibility of contamination from grease or organic/water vapors from the operator's hands. It is advisable that a spare detector is purchased with the instrument, in order to be fully prepared for any unforeseen circumstances.

Turbomolecular Pumps

The number of turbomolecular pumps used in modern ICP-MS systems will depend on the design of the mass spectrometer. Some of the newer instruments are using a single, twin-throated turbomolecular pump. It is too early to assess the reliability of this design. However, most of the instruments running today use two turbo pumps to create the operating vacuum for the main mass analyzer/detector chamber and the ion optic region. The lifetime of these pumps is dependant on a number of factors, including the pumping capacity of the turbo pump (L/sec), the size (volume) of the vacuum chamber to be pumped, the orifice diameter of the interface cones, and the time the instrument is running. While some instruments are still using the same turbo pumps after 5–10 years of operation, the normal lifetime of a pump in an instrument that has a reasonably high sample workload is in the order of 2–3 years. This is an approximation and will obviously vary depending on the make and design of the pump (especially the type of bearings that are used). As the turbo pump is one of the most expensive components of an ICP-MS system, this should be factored into the overall running costs of the instrument over its operating lifetime.

It is worth pointing out that although the turbo pump is not generally considered a part of routine maintenance, most instruments use a "Penning" (or similar) gauge to monitor the vacuum in the main chamber. Unfortunately, this gauge can become dirty over time and lose its ability to measure the correct pressure. The frequency of this is almost impossible to predict but is closely related to the types and numbers of samples analyzed. A dirty Penning gauge can show itself in a number of ways, but usually a sudden drop in pressure or fluctuations in the signal are two of the most common indications. When these happen, the gauge must be removed and cleaned. This should be carried out by an experience operator or service engineer, because it is a fairly complicated procedure to remove the gauge, clean it, maintain the correct electrode geometry, and reinstall it correctly back into the instrument. It is also further complicated by the fact that a Penning gauge is operated at high voltage.

Mass Analyzer

Under normal circumstances, there is no need for the operator to be concerned about routine maintenance on the mass analyzer. With modern turbo-molecular pumping systems, it is highly unlikely there will be any pump or sample-related contamination problems associated with the quadrupole, magnetic sector, or time-of-flight mass analyzer. This certainly was not the case with some of the early instruments that used oil-based diffusion pumps, because many researchers experienced contamination of the quadrupole and prefilters by oil vapors from the pumps. Today, it is fairly common for turbo-molecular-based mass analyzers to require no maintenance of the quadrupole rods over the lifetime of the instrument, other than an inspection carried out by a service engineer on an annual basis. However, in extreme cases, particularly with older instruments, it might require removal and cleaning of the quadrupole assembly, in order to get acceptable peak resolution and abundance sensitivity performance.

I think the overriding message I would like to leave you with in this chapter is that routine maintenance cannot be over emphasized in ICP-MS. Although it might be considered a mundane and time-consuming chore, it can have a significant impact on the "up-time" of your instrument. Read the routine maintenance section of the operators' manual and understand what is required. It is essential that time is scheduled on a weekly, monthly, and quarterly basis to carry out preventative maintenance on your instrument. In addition, you should budget for an annual preventative maintenance contract, where the service engineer checks out all the important instrumental components and systems on a regular basis to make sure they are all working correctly. This might not be as critical if you work in an academic environment, where the instrument might be used for extended periods, but in my opinion, is absolutely critical if you are a commercial laboratory that is using the instrument to generate revenue. There is no question that spending the time to keep your ICP-MS in good working order can mean the difference between owning an instrument whose performance could be slowly degrading without your knowledge or one that is always working in "peak" condition.

17

Alternate Sampling Accessories

Today, nonstandard sampling tools such as laser ablation systems, flow injection analyzers, autodilutors, electrothermal vaporizers, desolvation equipment, direct injection nebulizers, and chromatography separation devices are considered critical to enhance the practical capabilities of ICP mass spectrometry for real-world samples. Because they were developed over 10 years ago, these kinds of alternate sampling accessories have proved to be invaluable for certain applications that are considered problematic for ICP-MS.

It is recognized that standard ICP-MS instrumentation using a traditional sample introduction system composed of a spray chamber and nebulizer has certain limitations, particularly when it comes to the analysis of complex samples. Some of these known limitations include

- Total dissolved solids must be kept below 0.2%.
- Long washout times required for samples with a heavy matrix.
- Sample throughput is limited by the sample introduction process.
- Contamination issues with samples requiring multiple sample preparation steps.
- Dilutions and addition of internal standards can be labor intensive and time-consuming.
- If matrix has to be removed, it has to be done off-line.
- Matrix suppression can be quite severe with some samples.
- Matrix components can generate severe spectral overlaps on analytes.
- Organic solvents can present unique problems.
- The analysis of solids and slurries is very difficult.
- Not suitable for the analysis of elemental species or oxidation states.

Such were the demands of real-world users to overcome these kinds of problem areas that instrument companies devised different strategies based on the type of samples being analyzed. Some of these strategies involved parameter optimization or the modification of instrument components, but it was clear that this approach alone was not going to solve every conceivable problem. For this reason, they turned their attention to the development of sampling accessories, which were optimized for a particular application problem or sample type. Over the past 10–15 years, this demand has led to the commercialization of specialized sample introduction tools—not only by the instrument manufacturers themselves, but also by companies specializing in these kinds of accessories. The most common ones used today include:

- Laser ablation/sampling (LA/S)
- Flow injection analysis (FIA)
- Electrothermal vaporization (ETV)
- Desolvation systems
- Direct injection nebulizers (DIN)
- Chromatography separation techniques

Let us now take a closer look at each of these techniques to understand their basic principles and what benefits they bring to ICP-MS.

LASER ABLATION/SAMPLING

The limitation of ICP-MS to analyze solid materials (without the need for wet chemical dissolution/digestion methods) led to the development of laser ablation. The principle behind this approach is the use of a high-powered laser to ablate the surface of a solid and sweep the sample aerosol into the ICP mass spectrometer for analysis in the conventional way (1).

Before we go on to describe some typical applications suited to laser ablation ICP-MS, let us first take a brief look at the history of analytical lasers and how they eventually became such a useful sampling tool. The use of lasers as vaporization devices was first investigated in the early 1960s. When light energy with an extremely high power density interacts with a solid material, the photon-induced energy is converted into thermal energy, resulting in vaporization and removal of the material from the surface of the solid (2). Some of the early researchers used ruby lasers to induce a plasma discharge on the surface of the sample and measure the emitted light with an atomic emission spectrometer (3). Although this proved useful for certain applications, the technique suffered from low sensitivity, poor precision, and severe matrix effects caused by nonreproducible excitation characteristics. Over the years, various improvements were made to this basic design with very little success

(4), because the sampling process and the ionization/excitation process (both under vacuum) were still intimately connected and highly interactive with each other.

This limitation led to the development of laser ablation as a sampling device for atomic spectroscopy instrumentation, where the sampling step was completely separated from the excitation or ionization step. The major benefit being that each step could be independently controlled and optimized. These early devices used a high-energy laser to ablate the surface of a solid sample and the resulting aerosol swept into some kind of atomic spectrometer for analysis. Although initially used with atomic absorption (5,6) and plasma-based emission techniques (7,8), it was not until the mid-1980s when lasers were coupled with ICP-MS that the analytical community stood up and took notice (9). For the first time researchers were showing evidence that virtually any type of solid could be vaporized, irrespective of electrical characteristics, surface topography, size or shape, and transported into the ICP for analysis by atomic emission or mass spectrometry. This was an exciting breakthrough for ICP-MS, because it meant the technique could be used for the bulk sampling of solids, or if required, for the analysis of small spots and/or microinclusions, in addition to being used for the analysis of solutions.

Commercial Systems for ICP-MS

The first laser ablation systems developed for ICP instrumentation were based on solid-state ruby lasers, operating at 694 nm. These were developed in the early 1980s but did not prove to be successful for a number of reasons, including poor stability, low power density, low repetition rate, and large beam diameter, which made them limited in their scope and flexibility as a sample introduction device for trace element analysis. It was at least another 5 years before any commercial instrumentation became available. These early commercial laser ablation systems, which were specifically developed for ICP-MS, used the neodymium-doped yttrium aluminum garnet (Nd:YAG) design, operating at the primary wavelength of 1064 nm—in the infrared (10). They initially showed a great deal of promise because analysts were finally able to determine trace levels directly in the solid without sample dissolution. However, it soon became apparent that they did not meet the expectations of the analytical community, for many reasons including complex ablation characteristics, poor precision, not optimized for microanalysis, and, because of poor laser coupling, were unsuitable for many types of solids. By the early 1990s, most of the laser ablation systems purchased were viewed as novel and interesting but not suited to solve real-world application problems.

These basic limitations in IR laser technology led researchers to investigate the benefits of shorter wavelengths. Systems were developed that were based on Nd:YAG technology at the 1064-nm primary wavelength, but utilizing optical components to double (532 nm), quadruple (266 nm), and quintuple (213 nm) the frequency. Innovations in lasing materials and electronic design together with better thermally characteristics produced higher energy with higher pulse-to-pulse stability. These more advanced UV lasers showed significant improvements, particularly in the area of coupling efficiency, making them more suitable for a wider array of sample types. In addition, the use of higher quality optics allowed for a more homogeneous laser beam profile, which provided the optimum energy density to couple with the sample matrix. This resulted in the ability to make spots much smaller and with more controlled ablations irrespective of sample material, which were critical for the analysis of surface defects, spots, and microinclusions

Excimer Lasers

The successful trend toward shorter wavelengths and the improvements in the quality of optical components also drove the development of UV gas-filled lasers, such as XeCl (308 nm), KrF (248 nm), and ArF (193 nm) excimer lasers. These showed great promise, especially the ones operated at shorter wavelengths that were specifically designed for ICP-MS. Unfortunately, they necessitated a more sophisticated beam delivery system, which tended to make them more expensive. In addition, the complex nature of the optics and the fact that gases had to changed on a routine basis made them a little more difficult to use and maintain and, as a result, required a more skilled operator to run them. However, their complexity was far outweighed by their better absorption capabilities for UV transparent materials (such as calcites, fluorites, and silicates), smaller particle size, and higher flow of ablated material. There was also evidence to suggest that the shorter wavelength excimer lasers exhibit better elemental fractionation characteristics (typically defined as the intensity of certain elements varying with time, relative to the dry aerosol volume) than the longer wavelength Nd:YAG design, because they produce smaller particles that are easier to volatilize.

Benefits of Laser Ablation for ICP-MS

Today, there are a number of commercial laser ablation designs on the market today including 266- and 213-nm Nd:YAG and 193-nm ArF excimer lasers. They all have varying output energy, power density, and beam profiles and although each one has different ablation characteristics, they all work extremely well depending on the types of samples being analyzed and the data quality requirements. Laser ablation is now considered a very reliable sampling technique for ICP-MS, which is capable of producing data of the very highest quality directly on solid samples and powders. Some of the many benefits offered by this technique include:

- Direct analysis of solids without dissolution.
- Ability to analyze virtually any kind of solid material including rocks, minerals, metals, ceramics, polymers, plastics, plant material, and biological specimens.
- Ability to analyze a wide variety of powders by pelletizing with a binding agent.
- No requirement for sample to be electrically conductive.
- Sensitivity in the ppb to ppt range, directly in the solid.
- Labor-intensive sample preparation steps are eliminated, especially for samples such as plastics and ceramics that are extremely difficult to get into solution.
- Contamination is minimized because there are no digestion/dilution steps.
- Reduced polyatomic spectral interferences compared to solution nebulization.
- Examination of small spots, inclusions, defects, or microfeatures on the surface of sample.
- Elemental mapping across the surface of a mineral.
- Depth profiling to characterize thin films or coatings.

Let us now take a closer look at the strengths and weaknesses of the different laser designs based on the application requirements.

Optimum Laser Design Based on Application Requirements

The commercial success of laser ablation was initially driven by its ability to directly analyze solid materials such as rocks, minerals, ceramics, plastics, and metals, without going through a sample dissolution stage. Table 17.1 represents some typical multielement detection limits in NIST 612 glass generated with a 266-nm Nd:YAG design (11). It can be seen that, for most of the elements, sub-ppb detection limits in the solid material are achievable. This kind of performance is typically obtained using larger spot sizes in the order of 100 μm in diameter, which is ideally suited to 266-nm laser technology. However, the desire for ultra trace analysis of optically challenging materials, such as calcite, quartz, glass, and fluorite, combined with the capability to characterize small spots and microinclusions, proved very challenging for the 266-nm design. The major reason being that the ablation process is less controlled and as a result it is difficult to ablate a minute area without removing some of the surrounding material. In addition, erratic ablating of

TABLE 17.1   Typical Detection Limits Achievable in NIST 612 SRM Glass
Using a 266-nm Nd:YAG Laser Ablation System Coupled to an ICP Mass
Spectrometer

Element	3σ DLs (ppb)	Element	3σ DLs (ppb)
B	3.0	Ce	0.05
Sc	3.4	Pr	0.05
Ti	9.1	Nd	0.5
V	0.4	Sm	0.1
Fe	13.6	Eu	0.1
Co	0.05	Gd	1.5
Ni	0.7	Dy	0.5
Ga	0.2	Ho	0.01
Rb	0.1	Er	0.2
Sr	0.07	Yb	0.4
Y	0.04	Lu	0.04
Zr	0.2	Hf	0.4
Nb	0.5	Ta	0.1
Cs	0.2	Th	0.02
Ba	0.04	U	0.02
La	0.05		

Source: Courtesy of Cetac Technologies.

the sample initially generates larger particles, which are not fully ionized in the plasma and therefore contribute to poor precision (12). Although modifications helped to improve ablation behavior, it was not totally successful, because of the basic limitation of the 266-nm laser to couple efficiently to UV transparent materials. The weaknesses in 266-nm technology eventually led to the development of 213-nm lasers (13) because of the recognized superiority of shorter wavelengths to exhibit a higher degree of absorbance in transparent materials (14).

Analytical chemists, particularly in the geochemical community, welcomed 213-nm UV lasers with great enthusiasm, because they now had a sampling tool, which offered much better control of the ablation process, even for easily fractured minerals. This is demonstrated in Figure 17.1, which shows the difference between 266- and 213-nm ablation craters in a sample of apatite (a fluoride/phosphate matrix found in human teeth). It can be seen that the craters produced with the 213-nm laser are relatively round and symmetrical, whereas the 266-nm craters are irregular and show ablated material around the sides of the craters.

This significant difference in crater geometry between the two systems is translated into a difference in the rate of depth penetration, size distribution,

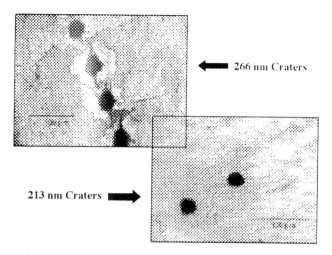

266 nm Craters

213 nm Craters

FIGURE 17.1 Craters produced with the 213-nm laser system are relatively round and symmetrical, whereas craters produced using the 266-nm are more irregular and show excess ablated material around the sides of the craters. (Courtesy of New Wave Research.)

and volume of particles reaching the plasma. With the 266-nm laser system, a high volume burst of material is initially observed producing a spike in the signal whereas with the 213-nm laser, the signal gradually increases and levels off quickly, indicating a more consistent stream of small particles being delivered to the plasma. Therefore when analyzing this type of mineral with the 266-nm design, it is sometimes necessary to filter out the first 100 to 200 shots of the ablation process, to ensure that no data are taken during the initial burst of material—which might be problematic when analyzing small spots or inclusions.

The benefits of 213-nm lasers emphasize that matrix independence, high spatial resolution, and the ability to couple with UV transparent materials without fracturing (particularly for small spots or depth analysis studies) were very important for geochemical-type applications. These findings led researchers to study even shorter wavelengths and in particular 193-nm ArF excimer technology. Besides their accepted superiority in coupling efficiency, a major advantage of the 193-nm design is that it utilizes a fundamental wavelength and therefore achieves much higher energy transfer, compared to a Nd:YAG solid-state system that utilizes crystals to quadruple or quintuple the frequency. Additionally, the less coherent nature of the excimer beam enables better optical homogenization resulting in an even flatter beam pro-

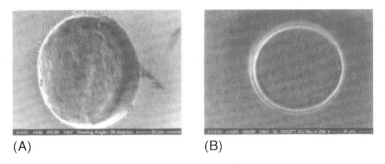

(A)   (B)

FIGURE 17.2 Scanning electron microscope (SEM) images of a 213-nm Nd:YAG laser crater on the left (A) and an optically homogenized flat beam 193-nm ArF excimer laser crater on the right (B). (Courtesy of New Wave Research.)

file. The overall benefit is that cleaner, flatter craters are produced down to approximately 3–4 μm in diameter. This provides far better control of the ablation process, which is especially important for depth profiling and fluid inclusion analysis. This is demonstrated in Figure 17.2, which shows scanning electron microscope (SEM) images of a sample of glass ablated with a 213-nm Nd:YAG laser on the left (A) and a crater ablated with a flat beam 193-nm

Material	193nm	213nm	266nm		Legend	
					Bulk	□
					Small feature	✱
Processed metals steel, zinc, copper, alloys, noble metals, etc.	⊗	□ = ✱	□○ [icon]		Fluid inclusion	⊙
					UV transparent	≋
Forensics glass, hair, ceramics, diamonds, etc.	✱ = ≋	□ ✱ = ≋	□○		Spatial resolution	≡
					Liquid	♦
paper, plastic, ballistics	□ ✱ =	□ ✱ =	□○ [icon]		Large spot > 350μ	○
Geochemistry zircons, fluid inclusions, calcite, fluorite, etc	✱ = ≋ ⊙	□ ✱ = ≋ ⊙	⊗		Recommended	
Shells/Carbonates otoliths, forams, mollusks, manganese nodules, etc.	□ ✱ =	□ ✱ =	⊗		Alternative	
Petrology oil, coal, basalts, etc.	□ ✱ = ♦	□ ✱ = ♦	□♦○ [icon]		Not Recommended	
Biological tissues, gels, tree rings, etc.	□ ✱ = ♦	□ ✱ = ♦	□♦○ [icon]		ICP-OES Friendly	
Environmental air filters, wear metals filters, soils, etc.	⊗	□ ✱ =	□○			
Nuclear vitreous waste, environmental clean-up, etc.	□ ✱ =	□ ✱ =	□○			

FIGURE 17.3 Some broad guidelines as to the optimum laser wavelength to use, based on the sample material. (Courtesy of New Wave Research.)

ArF excimer laser using an internally homogenized beam delivery system on the right (B). It can be seen that the excimer laser produces a much flatter and smoother crater than the Nd:YAG laser system (15).

The benefits of laser ablation system are now fairly well documented by the large number of application references in the public domain, which describe the analysis of metals, ceramics, polymers, rocks, minerals, biological tissue, and many other sample types (16–21). These references should be investigated further to better understand the optimum configuration, design, and wavelength of laser ablation equipment for different types of sample matrices. Figure 17.3 gives some broad guidelines of which design to use based on the applications being carried out. It should be emphasized that there are many overlapping areas when selecting the optimum laser system for the sample type, so this table should mainly be used for comparison purposes and not as a definitive guide.

FLOW INJECTION ANALYSIS

Flow injection (FI) is a powerful front-end sampling accessory for ICP-MS that can be used for preparation, pretreatment, and delivery of the sample. Originally described by Ruzicka and Hansen (22), flow injection involves the introduction of a discrete sample aliquot into a flowing carrier stream. Using a series of automated pumps and valves, procedures can be carried out on-line to physically or chemically change the sample or analyte, before introduction into the mass spectrometer for detection. There are many benefits of coupling flow injection procedures to ICP-MS, including:

- Automation of on-line sampling procedures, including dilution and additions of reagents.
- Minimum sample handling translates into less chance of sample contamination.
- Ability to introduce low sample/reagent volumes.
- Improved stability with harsh matrices.
- Extremely high sample throughput using multiple loops.

In its simplest form, FI-ICP-MS consists of a series of pumps and an injection valve preceding the sample introduction system of the ICP mass spectrometer. A typical manifold used for microsampling is shown in Figure 17.4.

In the fill position, the valve is filled with the sample. In the inject position, the sample is swept from the valve and carried to the ICP by means of a carrier stream. The measurement is usually a transient profile of signal versus time, as shown by the signal profile in Figure 17.4. The area of the signal

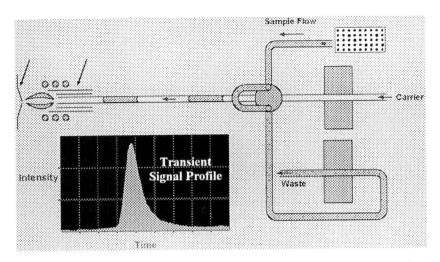

FIGURE 17.4 Schematic of a flow injection system used for the process of microsampling.

profile measured is greater for larger injection volumes, but for volumes of 500 μL or greater, the signal peak height reaches a maximum equal to that observed using continuous solution aspiration. The length of a transient peak in flow injection is typically 20–60 sec, depending on the size of the loop. This means if multielement determinations are a requirement, all the data quality objectives for the analysis, including detection limits, precision, dynamic range, and number of elements, etc., must be achieved in this time frame. Similar to laser ablation, if a sequential mass analyzer such as a quadrupole or single collector magnetic sector system is used, the electronic scanning, dwelling, and settling times must be optimized in order to capture the maximum amount of multielement data in the duration of the transient event (23). This can be seen in greater detail in Figure 17.5, which shows a 3D transient plot of intensity versus mass in the time domain, for the determination of a group of elements.

Some of the many on-line procedures that are applicable to FI-ICP-MS include:

Microsampling for improved stability with heavy matrices (24)
Automatic dilution of samples/standards (25)
Standards addition (26)
Cold vapor and hydride generation for enhanced detection capability for elements such as Hg, As, Sb, Bi, Te, and Se (27)

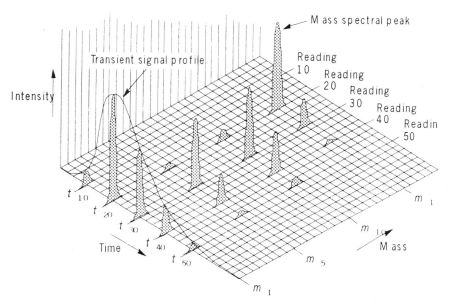

FIGURE 17.5 A 3D plot of intensity versus mass in the time domain, for the determination of a group of elements in a transient peak. (Courtesy of PerkinElmer Life and Analytical Sciences.)

Matrix separation and analyte preconcentration using ion exchange procedures (28)

Elemental speciation (29)

Flow injection coupled to ICP-MS has shown itself to be very diverse and flexible in meeting the demands presented by complex samples as indicated in the above references. However, one of the most exciting areas of research at the moment is in the direct analysis of seawater by flow injection ICP-MS. Traditionally, the analysis of seawater is very difficult by ICP-MS, because of two major problems. First, the high NaCl content will block the sampler cone orifice over time, unless a 10–20-fold dilution is made of the sample. This is not such a major problem with coastal waters, because the levels are high enough. However, if the sample is open ocean seawater, this is not an option because the trace metals are at a much lower level. The other difficulty associated with the analysis of seawater is that ions from the water, chloride matrix, and the plasma gas can combine to generate polyatomic spectral interferences, which are a problem, particularly for the first-row transition metals.

Attempts have been made over the years to remove the NaCl matrix and preconcentrate the analytes using various types of chromatography and ion exchange column technology. One such early approach was to use an HPLC system coupled to an ICP mass spectrometer utilizing a column packed with silica immobilized 8-hydroxyquinoline (30). This worked reasonably well but was not considered a routine method, because silica-immobilized 8-hydroxyquinoline was not commercially available and also spectral interferences produced by HCl and HNO$_3$ (used to elute the analytes) precluded the determination of a number of the elements, such as Cu, As, and V. More recently, chelating agents based on the iminodiacetate acid functionality group have gained wider success but are still not considered truly routine for a number of reasons, including the necessity for calibration using standard additions, the requirement of large volumes of buffer to wash the column after loading the sample, and the need for conditioning between samples because some ion exchange resins swell with changes in pH (31–33).

However, a research group at the NRC in Canada has developed a very practical on-line approach, using a flow injection sampling system coupled to an ICP mass spectrometer (28). Using a special formulation of a commercially available, iminodiacetate ion exchange resin (with a macroporous meth-

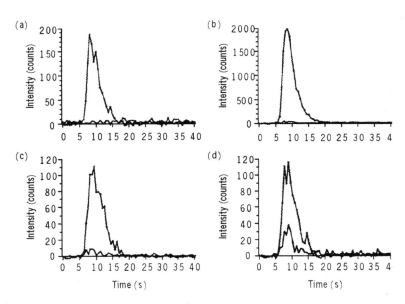

FIGURE 17.6   Analyte and blank spectral scans of (a) Co, (b) Cu, (c) Cd, and (d) Pb in NASS-4 open-ocean seawater certified reference material, using flow injection coupled to ICP-MS. (From Ref. 28.)

TABLE 17.2 Analytical Results for NASS-4 Open-Ocean Seawater Certified
Reference Material, Using Flow Injection ICP-MS Methodology

| | | NASS-4 (ppb) | |
Isotope	LOD (ppt)	Determined	Certified
$^{51}V^+$	4.3	1.20 ± 0.04	Not certified
$^{63}Cu^+$	1.2	0.210 ± 0.008	0.228 ± 0.011
$^{60}Ni^+$	5	0.227 ± 0.027	0.228 ± 0.009
$^{66}Zn^+$	9	0.139 ± 0.017	0.115 ± 0.018
$^{55}Mn^+$	Not reported	0.338 ± 0.023	0.380 ± 0.023
$^{59}Co^+$	0.5	0.0086 ± 0.0011	0.009 ± 0.001
$^{208}Pb^+$	1.2	0.0090 ± 0.0014	0.013 ± 0.005
$^{114}Cd^+$	0.7	0.0149 ± 0.0014	0.016 ± 0.003

Source: From Ref. 28.

acrylate backbone), trace elements can be separated from the high concentrations of matrix components in the seawater, with a pH 5.2 buffered solution. The trace metals are subsequently eluted into the plasma with 1 M HNO_3, after the column has been washed out with deionized water. The column material has sufficient selectivity and capacity to allow accurate determinations at parts-per-trillion levels using simple aqueous standards, even for elements such as V and Cu, which are notoriously difficult in a chloride matrix. This can be seen in Figure 17.6, which shows spectral scans for a selected group of elements in a certified reference material open-ocean seawater sample (NASS-4), and Table 17.2, which compares the results for this methodology with the certified values, together with the limits of detection (LOD). Using this on-line method, the turnaround time is less than 4 min per sample, which is considerably faster than other high-pressure chelation techniques reported in the literature.

ELECTROTHERMAL VAPORIZATION

Electrothermal atomization (ETA) for use with atomic absorption (AA) has proven to be a very sensitive technique for trace element analysis over the last three decades. However, the possibility of using the atomization/heating device for electrothermal vaporization (ETV) sample introduction into an ICP mass spectrometer was identified in the late 1980s (34). The ETV sampling process relies on the basic principle that a carbon furnace or metal filament can be used to thermally separate the analytes from the matrix components and then sweep them into the ICP mass spectrometer for analysis.

This is achieved by injecting a small amount of the sample (usually 20–50 μL via an autosampler) into a graphite tube or onto a metal filament. After the sample is introduced, drying, charring, and vaporization are achieved by slowly heating of the graphite tube/metal filament. The sample material is vaporized into a flowing stream of carrier gas, which passes through the furnace or over the filament during the heating cycle. The analyte vapor recondenses in the carrier gas and is then swept into the plasma for ionization.

One of the attractive characteristics of ETV for ICP-MS is that the vaporization and ionization steps are carried out separately, which allows for the optimization of each process. This is particularly true when a heated graphite tube is used as the vaporization device, because the analyst typically has more control of the heating process and as a result can modify the sample by means of a very precise thermal program before it is introduced to the ICP for ionization. By boiling off and sweeping the solvent and volatile matrix components out of the graphite tube, spectral interferences arising from the sample matrix can be reduced or eliminated. The ETV sampling process consists of six discrete stages: sample introduction, drying, charring (matrix removal), vaporization, condensation, and transport. Once the sample has been introduced, the graphite tube is slowly heated to drive off the solvent. Opposed gas flows, entering from each end of the graphite tube, then purge the sample cell by forcing the evolving vapors out the dosing hole. As the temperature increases, volatile matrix components are vented during the charring steps. Just prior to vaporization, the gas flows within the sample cell are changed. The central channel (nebulizer) gas then enters from one end of the furnace, passes through the tube, and exits out the other end. The sample-dosing hole is then automatically closed, usually by means of a graphite tip, to ensure no analyte vapors escape. After this gas flow pattern has been established, the temperature of the graphite tube is ramped up very quickly, vaporizing the residual components of the sample. The vaporized analytes either recondense in the rapidly moving gas stream or remain in the vapor phase. These particulates and vapors are then transported to the ICP in the carrier gas where they are ionized by the ICP for analysis in the mass spectrometer.

Another benefit of decoupling the sampling and ionization processes is the opportunity for chemical modification of the sample. The graphite furnace itself can serve as a high temperature reaction vessel where the chemical nature of compounds within it can be altered. In a manner similar to that used in atomic absorption, chemical modifiers can change the volatility of species to enhance matrix removal and/or increase elemental sensitivity (35). An alternate gas such as oxygen may also be introduced into the sample cell to aid in the charring of the carbon in organic matrices such as biological or petrochemical samples. Here the organically bound carbon reacts with the oxygen gas to produce CO_2, which is then vented from the system. A typical

ETV sampling device, showing the two major steps of sample pretreatment (drying and ashing) and vaporization into the plasma, is seen schematically in Figure 17.7.

Over the past 15 years, ETV sampling for ICP-MS has mainly been used for the analysis of complex matrices including geological materials (36), biological fluids (37), seawater (38), and coal slurries (39), which have proven difficult or impossible by conventional nebulization. By removal of the matrix components, the potential for severe spectral and matrix-induced interferences is dramatically reduced. Although ETV-ICP-MS was initially applied to the analysis of very small sample volumes, the advent of low-flow nebulizers has mainly precluded its use for this type of work.

An example of the benefit of ETV sampling is in the analysis of samples containing high concentrations of mineral acids such as HCl, HNO_3, and

(a)

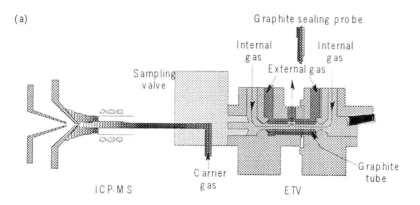

(b)

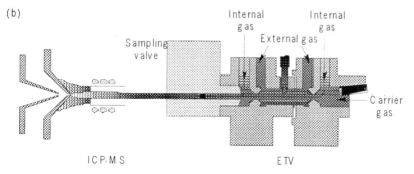

FIGURE 17.7 A graphite furnace ETV sampling device for ICP-MS, showing the two distinct steps of sample pretreatment and vaporization into the plasma. (Courtesy of PerkinElmer Life and Analytical Sciences.)

H_2SO_4. Besides physically suppressing analyte signals, these acids generate massive polyatomic spectral overlaps, which interfere with many analytes including As, V, Fe, K, Si, Zn, and Ti. By carefully removing the matrix components with the ETV device, the determination of these elements becomes relatively straightforward. This is exemplified in Figure 17.8, which shows a spectral display in the time domain for 50-pg spikes of a selected group of elements in concentrated hydrochloric acid (37% w/w) using a graphite furnace-based ETV-ICP-MS (40). It can be seen in particular that good sensitivity is obtained for $^{51}V^+$, $^{56}Fe^+$, $^{75}As^+$, which would have been virtually impossible by direct aspiration because of spectral overlaps from $^{39}ArH^+$, $^{35}Cl^{16}O^+$, $^{40}Ar^{16}O^+$, and $^{40}Ar^{35}Cl^+$, respectively. The removal of the chloride and water from the matrix translates into ppt detection limits directly in 37% HCl, as shown in Table 17.3.

It can also be seen in Figure 17.8 that the elements are vaporized off the graphite tube in order of their boiling points. In other words, antimony and magnesium, which are the most volatile, are driven off first, while V and Mo, which are the most refractory, come off last. However, although they emerge at different times, the complete transient event lasts less than 3 sec. This

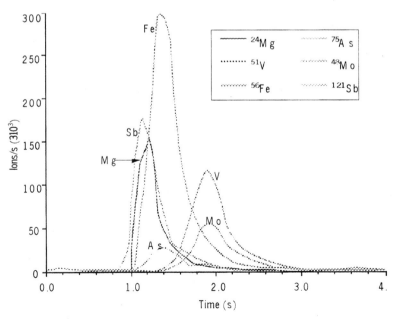

FIGURE 17.8 A temporal display of 50 pg of Mg, Sb, As, Fe, V, and Mo in 37% hydrochloric acid by ETV-ICP-MS. (From Ref. 40.)

TABLE 17.3 Detection Limits for V, Fe, and As in 37% Hydrochloric Acid by ETV-ICP-MS

Element	DL (ppt)
$^{51}V^+$	50
$^{56}Fe^+$	20
$^{75}As^+$	40

Source: From Ref. 40.

physical time limitation, imposed by the duration of the transient signal, makes it imperative that all isotopes of interest be measured under the highest signal-to-noise conditions throughout the entire event. The rapid nature of the transient has also limited the usefulness of ETV sampling for routine multielement analysis, because realistically only a small number of elements can be quantified with good accuracy and precision in less than 3 sec. In addition, the development of low flow nebulizers, desolvation devices, and collision cell technology has meant that rapid multielement analysis can now be carried out on difficult samples without the need for ETV sample introduction.

DESOLVATION DEVICES

Desolvation devices are mainly used in ICP-MS to reduce the amount of solvent entering the plasma. With organic samples, desolvation is absolutely critical, because most volatile solvents would extinguish the plasma if they were not removed or at least significantly reduced. However, desolvation of most types of samples can be very useful because it reduces the severity of the solvent-induced spectral interferences such as oxides, hydroxides, and argon/solvent-based polyatomics that are common in ICP-MS. The most common desolvation systems used today include:

- Water-cooled spray chambers
- Peltier-cooled spray chambers
- Ultrasonic nebulizers (USN) with water/peltier coolers
- Ultrasonic nebulizers (USN) with membrane desolvation
- Microconcentric nebulizers (MCN) with membrane desolvation

Let us take a closer look at these devices.

Cooled Spray Chambers

Water- and/or peltier (thermoelectric)-cooled spray chambers are standard on a number of commercial instruments. They are usually used with conventional or low flow pneumatic nebulizers to reduce the amount of solvent entering the plasma. This has the effect of minimizing solvent-based spectral interferences formed in the plasma and can also help to reduce the effects of a nebulizer-flow-induced secondary discharge at the interface of the plasma with the sampler cone. With some organic samples, it has proved to be very beneficial to cool the spray chamber to $-10°C$ to $-20°C$ (with an ethylene glycol mix) in addition to adding a small amount of oxygen into the nebulizer gas flow. This has the effect of reducing the amount of organic solvent entering the interface, which is beneficial in eliminating the build-up of carbon deposits on the sampler cone orifice and also minimizing the problematic carbon-based spectral interferences (41).

Ultrasonic Nebulizers

Ultrasonic nebulization was first developed in the late 1980s for use with ICP optical emission (42). Its major benefit was that it offered an approximately $10\times$ improvement in detection limits, because of its more efficient aerosol generation. However, this was not such an obvious benefit for ICP-MS, because more matrix entered the system compared to a conventional nebulizer, increasing the potential for signal drift, matrix suppression, and spectral interferences. This was not such a major problem for simple aqueous samples but was problematic for real-world matrices. The elements that showed the most improvement were the ones that benefited from lower solvent-based spectral interferences. Unfortunately, many of the other elements exhibited higher background levels and as a result showed no significant improvement in detection limit. In addition, because of the increased amount of matrix entering the mass spectrometer, it usually necessitated the need for larger dilutions of the sample, which again negated the benefit of using an USN with ICP-MS. This limitation led to the development of an ultrasonic nebulizer fitted with an additional membrane desolvator. This design virtually removed all the solvent from the sample, which dramatically improved detection limits for a large number of the problematic elements and also lowered oxide levels by at least an order of magnitude (43).

The principle of aerosol generation using an ultrasonic nebulizer is based on a sample being pumped onto a quartz plate of a piezo-electric transducer. Electrical energy of 1–2-MHz frequency is coupled to the transducer, which causes it to vibrate at high frequency. These vibrations disperse the sample into a fine droplet aerosol, which is carried in a stream of argon. With a conventional ultrasonic nebulizer, the aerosol is passed through a heating

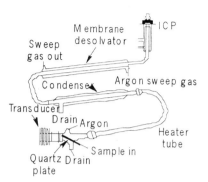

FIGURE 17.9 Schematic of an ultrasonic nebulizer fitted with a membrane desolvation system. (Courtesy of CETAC Technologies.)

tube and a cooling chamber where most of the sample solvent is removed as a condensate before it enters the plasma. If a membrane desolvation system is fitted to the ultrasonic nebulizer, it is positioned after the cooling unit. The sample aerosol enters the membrane desolvator, where the remaining solvent vapor passes through the walls of a tubular microporous PTFE membrane. A flow of argon gas removes the volatile vapor from the exterior of the membrane, while the analyte aerosol remains inside the tube and is carried into the plasma for ionization. This can be seen more clearly in Figure 17.9, which shows a schematic of an ultrasonic nebulizer, and Figure 17.10, which exemplifies the principles of membrane desolvation.

For ICP-MS, the system is best operated with both desolvation stages working, although for less demanding ICP-OES analysis, the membrane stage can be bypassed if required. The power of the system when coupled to an ICP mass spectrometer can be seen in Table 17.4, which compares the

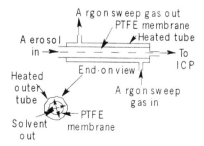

FIGURE 17.10 Principles of membrane desolvation. (Courtesy of CETAC Technologies.)

TABLE 17.4 Comparison of Sensitivity and Net Signal/Background Ratios Between a Crossflow Nebulizer and a Membrane Desolvation System

Analyte/BG	Mass (amu)	Crossflow nebulizer (cps)	Net analyte signal/BG	Membrane desolvation USN (cps)	Net analyte signal/BG
25 ppb $^{44}Ca^+$	44	2300	2300/7640	20,800	20,800/1730
(BG subtracted)			= 0.30		= 12.0
$^{12}C^{16}O_2^+$ (BG)		7640		1730	
10 ppb $^{56}Fe+$	56	95,400	95,400/868,000	262,000	262,000/8200
(BG subtracted)			= 0.11		= 32.0
$^{40}Ar^{16}O^+$ (BG)		868,000		8200	
10 ppb $^{57}Fe^+$	57	2590	2590/5300	6400	6400/200
(BG subtracted)			= 0.49		= 32.0
$^{40}Ar^{16}OH^+$ (BG)		5300		200	

Net analyte signal/BG is calculated as the background subtracted signal divided by the background.
Source: Courtesy of CETAC Technologies.

sensitivity (counts per second) and signal to background of a membrane desolvation USN with a conventional crossflow nebulizer for three classic solvent-based polyatomic interferences, $^{12}C^{16}O_2^+$ on $^{44}Ca^+$, $^{40}Ar^{16}O^+$ on $^{56}Fe^+$, and $^{40}Ar^{16}OH^+$ on $^{57}Fe^+$, using a quadrupole ICP-MS system. The sensitivities for the analyte isotopes are all background subtracted.

It can be seen that for all three analyte isotopes, the net signal-to-background ratio is significantly better with the membrane ultrasonic nebulizer than with the crossflow design, which is a direct impact of the reduction of the solvent-related spectral background levels. Although this approach works equally well and sometimes better when analyzing organic samples, it does not work for analytes that are bound to an organic molecule. The high volatility of certain types of organometallic species means that they could pass through the microporous Teflon membrane and never make it into the ICP-MS. In addition, samples with high dissolved solids, especially ones that are biological in nature, could possibly result in clogging the microporous membrane unless substantial dilutions are made. For these reasons, caution must be used when using a membrane desolvation system for the analysis of certain types of complex sample matrices.

Desolvating Microconcentric Nebulizers

A variation of the membrane desolvation system is with a microconcentric nebulizer in place of the ultrasonic nebulizer. A schematic of this design is shown in Figure 17.11.

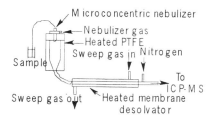

FIGURE 17.11 Schematic of a microconcentric nebulizer fitted with a membrane desolvation system. (Courtesy of CETAC Technologies.)

The benefit of this approach is not only the reduction in solvent-related spectral interferences with the membrane desolvation system, but also advantage can be taken of the microconcentric nebulizer's ability to aspirate very low sample volumes (typically 20–100 μL). This can be particularly useful, when sample volume is limited as in vapor phase decomposition (VPD) analysis of silicon wafers. The problem with this kind of demanding work is that there is typically only 500 μL of sample available, which makes it extremely difficult using a traditional low flow nebulizer, because it requires the use of both cool and normal plasma conditions to carry out a complete multielement analysis. By using an MCN with a membrane desolvation system, the full suite of elements, including the notoriously difficult ones such as Fe, K, and Ca, can be determined on 500 μL of sample using one set of normal plasma conditions (44).

It should be noted that conventional low flow nebulizers were described in greater detail in Chapter 3 on "Sample Introduction." The most common ones used in ICP-MS are based on the microconcentric design, which operate at 20–100 μL/min. Besides being ideal for small sample volumes, the major benefit of microconcentric nebulizers is that they are more efficient at producing small droplets than a conventional nebulizer. In addition, most low flow nebulizers use chemically inert plastic capillaries, which makes them well suited for the analysis of highly corrosive chemicals. This kind of flexibility has made low flow nebulizers very popular, particularly in the semiconductor industry where it is essential to analyze high-purity acids using a sample introduction system which is free of contamination (45).

DIRECT INJECTION NEBULIZERS

Direct injection nebulization is based on the principle of injecting a liquid sample under high pressure directly into the base of the plasma torch (60).

The benefit of this approach is that no spray chamber is required, which means that an extremely small volume of sample can be introduced directly into the ICP-MS with virtually no carryover or memory effects from the previous sample. Because they are capable of injecting <5 μL of liquid, they have found a use in applications where sample volume is limited or where the material is highly toxic or expensive.

They were initially developed over 10 years ago and found some success in certain niche applications such as introducing samples into an ICP-MS coupled to a chromatography separation devices or the determination of mercury by ICP-MS—which could not be adequately addressed by other nebulization systems. Unfortunately, they were not considered particularly user-friendly and as a result became less popular when other sample introduction devices were developed to handle microliter sample volumes. More recently, a refinement of the direct injection nebulizer has been developed called the direct inject high efficiency nebulizer (DIHEN), which appears to have overcome many of the limitations of the original design (61).

CHROMATOGRAPHIC SEPARATION DEVICES

ICP-MS has gained popularity over the years, based mainly on its ability to rapidly quantitate ultra trace metal contamination levels. However, in its basic design, ICP-MS cannot reveal anything about the metal's oxidation sate, alkylated form, or how it is bound to a biomolecule. The desire to understand in what form or species an element exists led researchers to investigate the combination of chromatographic separation devices with ICP-MS. The ICP mass spectrometer becomes a very sensitive detector for trace element speciation studies when coupled to a chromatographic separation device such as high performance liquid chromatography (HPLC), ion chromatography (IC), gas chromatography (GC), and capillary electrophoresis (CE). In these hybrid techniques, element species are separated based on their chromatograph retention/mobility times and then eluted/passed into the ICP mass spectrometer for detection (46). The intensity of the eluted peaks are then displayed for each isotopic mass of interest, in the time domain as shown in Figure 17.12, which shows a typical chromatogram for a selected group of masses between 60 and 75 amu.

There is no question that the extremely low detection capability of ICP-MS has allowed researchers in the environmental, biomedical, geochemical, and nutritional fields to gain a much better insight into the impact of different elemental species on us and our environment—something that would not have been possible 10–15 years ago. The majority of trace element speciation studies being carried out today can be broken down into three major cate-

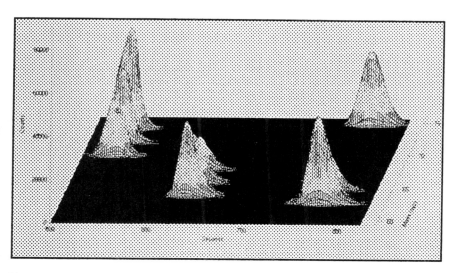

FIGURE 17.12 A typical chromatogram generated by a liquid chromatograph coupled to an ICP mass spectrometer, showing a temporal display of intensity against mass. (Courtesy of PerkinElmer Life and Analytical Sciences.)

gories—Redox systems, alkylated forms, and biomolecules. Let us take a closer look at these categories.

- **Redox** refers to reduction or oxidation of a metal, which changes its valency state. For example, hexavalent chromium, Cr (VI), is a powerful oxidant and extremely toxic but, in soils and water systems, reacts with organic matter to form trivalent chromium, Cr (III), which is the more common form of the element and is an essential micronutrient for plants and animals (47).
- **Alkylated forms.** Very often the natural form of an element can be toxic, while its alkylated form is relatively harmless—or vise versa. A good example of this is the element arsenic. Inorganic forms of the element such as As (III) and As (V) are toxic, whereas many of its alkylated forms such as monomethylarsonic acid (MMA) and dimethylarsonic acid (DMA) are relatively innocuous (48)
- **Metallo-biomolecules** are formed by the interaction of trace metals with complex biological molecules. For example, in animal studies, activity and mobility of an innocuous arsenic-based growth promoter are determined by studying its metabolic impact and excretion characteristics. So measurement of the biochemical form of arsenic is crucial in order to know its growth potential (49).

Table 17.5 represents a small cross section of speciation work that has been carried out by chromatography techniques coupled to ICP-MS in these three major categories.

As mentioned previously there is a large body of application work in the public domain that has investigated the use of different chromatographic separation devices, such as LC (50,51), IC (52), GC (53,54), and CE (55,56) with ICP-MS. A very popular area of research is the coupling of liquid chromatography systems (such as adsorption, ion-exchange, gel permeation, normal, or reverse-phase technology) with ICP-MS to gain valuable insight into the type of elemental species present in a sample. To get a better understanding of how the technique works, let us take a look at one of these applications—the determination of different forms of inorganic arsenic in soil, using ion-exchange HPLC coupled to ICP-MS.

Arsenic toxicity depends directly on the chemical form of the arsenic. In its inorganic form, arsenic is highly toxic while many of its organic forms are relatively harmless. Inorganic species of arsenic that are of toxological interest are the trivalent form [As (III)], such as arsenious acid, H_3AsO_3, and its arsenite salts; the pentavalent form [As (V)], such as arsenic acid, H_3AsO_5, and its arsenate salts; and arsine (AsH_3), a poisonous, unstable gas used in the manufacture of semiconductor devices. Arsenic is introduced into the environment and ecosystems from natural sources by volcanic activity and the weathering of minerals, and also from anthropogenic sources, such as ore smelting, coal burning, industrial discharge, and pesticide use. The ratio of natural arsenic to anthropogenic arsenic is approximately 60:40.

A recent study investigated a potential arsenic contamination of the soil in and around an industrial site. Soil in a field near the factory in question was sampled, as well as soil inside the factory grounds. The soil was dried, weighed, extracted with water, and filtered. This careful, gentle extraction procedure was used in order to avoid disturbing the distribution of arsenic

TABLE 17.5 Some Elemental Species That Have Been Studied by Researchers Using Chromatographic Separation Devices Coupled to ICP-MS

Redox systems	Alkylated forms	Biomolecules
Se (IV)/Se (VI)	Methyl-Hg, Ge, Sn, Pb, As, Sb, Se, Te, Zn, Cd, Cr	Organo-As, Se, Cd
As (III)/As (V)	Ethyl-Pb, Hg	Metallo-porphyrines
Sn (II)/Sn (IV)	Butyl-Sn	Metallo-proteins
Cr (III)/Cr (VI)	Phenyl-Sn	Metallo-drugs
Fe (II)/Fe (III)	Cyclohexyl-Sn	Metallo-enzymes

species originally present in the sample—an important consideration in speciation studies. Ten milliliters of sample was injected onto a column containing an amine-based anion exchange resin (Cetac Technologies—ANX 3206), where the different oxidation states of As were chromatographically extracted from the matrix and separated using a standard LC pump. The matrix components passed straight through the column, whereas the arsenic species were retained and then isochratically eluted into the nebulizer of the ICP mass spectrometer using 5 mM ammonium malonate. The arsenic species were then detected and quantified by running the instrument in the single-ion monitoring mode, set at mass 75—the only isotope for arsenic. This can be seen in greater detail in Figure 17.13, which shows that both As (III) and As (V) have been eluted off the column in less than 3 min using this HPLC-ICP-MS set-up. It can also be seen from the chromatogram that both species are approximately three orders of magnitude lower in the soil sample from the surrounding field, compared to the soil sample inside the factory grounds.

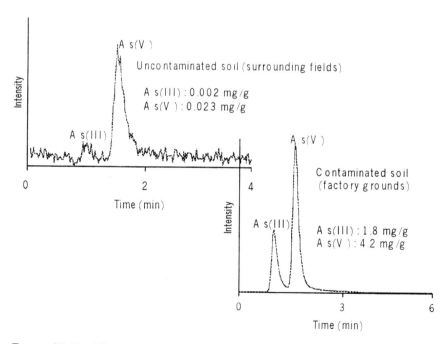

FIGURE 17.13 HPLC-ICP-MS chromatogram showing comparison of As (III) and As (V) levels in uncontaminated (left) and contaminated (right) soil samples in and around an industrial site. (Courtesy of Cetac Technologies.)

Although the arsenic does not exceed average global soil levels, it is a clear indication that the factory is a source of arsenic contamination.

It is worth mentioning that for some reverse-phase HPLC separations, gradient elution of the analyte species with mixtures of organic solvents such as methanol might have to be used. If this is a requirement, consideration must be given to the fact that large amounts of organic solvent will extinguish the plasma (57), so introduction of the eluent into the ICP mass spectrometer cannot be carried out using a conventional nebulization. For this reason, special sample introduction systems such as refrigerated spray chambers (58) or desolvation systems (59) have to be used, in addition to small amounts of oxygen in the sample aerosol flow to stop the build-up of carbon deposits on the sampler cone. Other approaches such as direct injection nebulization (60) have been used to introduce the sample eluent into the ICP-MS but historically have not gained widespread acceptance because of usability issues.

FURTHER READING

1. Denoyer ER, Fredeen KJ, Hager JW. Anal Chem 1991; 63(8):445–457A.
2. Ready JF. Affects of High Power Laser Radiation. New York: Academic Press, 1972. (Chapters 3–4).
3. Moenke-Blankenburg L. Laser Microanalysis. New York: Wiley, 1989.
4. Denoyer ER, Van Grieken R, Adams F, Natusch DFS. Anal Chem 1982; 54:26.
5. Carr JW, Horlick G. Spectrochim Acta 1982; 37B:1.
6. Kantor T, et al. Talanta 1979; 23:585.
7. Human HCG, et al. Analyst 1976; 106:265.
8. Thompson M, Goulter JE, Seiper F. Analyst 1981; 106:32.
9. Gray AL. Analyst 1985; 110:551.
10. Arrowsmith PA, Hughes SK. Appl Spectrosc 1988; 42:1231–1239.
11. Howe T, Shkolnik J, Thomas R. Spectroscopy 2001; 16(2):54–66.
12. Gunter D, Hattendorf B. Miner Assoc Can - Short Course Ser 2001; 29:83–91.
13. Jeffries TE, Jackson SE, Longerich HP. J Anal At Spectrom 1998; 13:935–940.
14. Russo RE, Mao XL, Haichen Borisov L. J Anal At Spectrom 2000; 15:1115–1120.
15. Russo RE, Mao XL, Gonzales J, Mao SS. Invited Talk. Winter Conference on Plasma Spectrochemistry, Scottsdale, AZ, 2002.
16. Jackson SE, Longerich HP, Dunning GR, Fryer BJ. Can Miner 1992; 30:1049–1064.
17. Gunther D, Heinrich CA. J Anal At Spectrom 1999; 14:1369.
18. Gunther D, Horn I, Hattendorf B. Fresenius J Anal Chem 2000; 368:4–14.
19. Wolf RE, Thomas C, Bohlke A. Appl Surf Sci 1998; 127–129:299–303.
20. Gonzalez J, Mao XL, Roy J, Mao SS, Russo RE. J Anal At Spectrom 2002; 17:1108–1113.
21. Jeffries TE, Perkins WT, Pearce NJG. Analyst 1995; 120:1365–1371.

22. Ruzicka J, Hansen EH. Anal Chim Acta 1975; 78:145.
23. Thomas R. Spectroscopy 2002; 17(5):54–66.
24. Stroh A, Voellkopf U, Denoyer E. J Anal At Spectrom 1992; 7:1201.
25. Israel Y, Lasztity A, Barnes RM. Analyst 1989; 114:1259.
26. Israel Y, Barnes RM. Analyst 1989; 114:843.
27. Powell MJ, Boomer DW, McVicars RJ. Anal Chem 1986; 58:2864.
28. Willie SN, Iida Y, McLaren JW. At Spectrosc 1998; 19(3):67.
29. Roehl R, Alforque MM. At Spectrosc 1990; 11(6):210.
30. McLaren JW, Lam JWH, Berman SS, Akatsuka K, Azeredo MA. J Anal At Spectrom 1993; 8:279–286.
31. Ebdon L, Fisher A, Handley H, Jones P. J Anal At Spectrom 1993; 8:979–981.
32. Taylor DB, Kingston HM, Nogay DJ, Koller D, Hutton R. J Anal At Spectrom 1996; 11:187–191.
33. Nelms SM, Greenway GM, Koller D. J Anal At Spectrom 1996; 11:907–912.
34. Park CJ, Van Loon JC, Arrowsmith P, French JB. Anal Chem 1987; 59:2191–2196.
35. Ediger RD, Beres SA. Spectrochim Acta 1992; 47B:907.
36. Park CJ, Hall GEM. J Anal At Spectrom 1987; 2:473–480.
37. Park CJ, Van Loon JC. Trace Elem Med 1990; 7:103.
38. Chapple G, Byrne JP. J Anal At Spectrom 1996; 11:549–553.
39. Voellkopf U, Paul M, Denoyer ER. Fresenius J Anal Chem 1992; 342:917–923.
40. Beres SA, Denoyer ER, Thomas R, Bruckner P. Spectroscopy 1994; 9(1):20–26.
41. McElroy F, Mennito A, Debrah E, Thomas R. Spectroscopy 1998; 13(2):42–53.
42. Olson KW, Haas WJ Jr, Fassel VA. Anal Chem 1977; 49(4):632–637.
43. Kunze J, Koelling S, Reich M, Wimmer MA. At Spectrosc 1998; 19:5.
44. Settembre G, Debrah E. Micro, June 1998; 16(6):79–84.
45. Aleksejczyk RA, Gibilisco D. Micro, September 1997.
46. Lobinski R, Pereiro IR, Chassaigne H, Wasik A, Szpunar J. J Anal At Spectrom 1998; 13:860–867.
47. Cox AG, McLeod CW. Mikrochim Acta 1992; 109:161–164.
48. Branch S, Ebdon L, O'Neill P. J Anal At Spectrom 1994; 9:33–37.
49. Dean JR, Ebdon L, Foulkes ME, Crews HM, Massey RC. J Anal At Spectrom 1994; 9:615–618.
50. Vela NP, Caruso JA. J Anal At Spectrom 1993; 8:787.
51. Caroli S, La Torre F, Petrucci F, Violante N. Environ: Sci Pollut Res 1994; 1(4):205–208.
52. Garcia-Alonso JI, Sanz-Medel A, Ebdon L. Anal Chim Acta 1993; 283:261–271.
53. Kim AW, Foulkes ME, Ebdon L, Hill SJ, Patience RL, Barwise AG, Rowland SJ. J Anal At Spectrom 1992; 7:1147–1149.
54. Hintelmann H, Evans RD, Villeneuve JY. J Anal At Spectrom 1995; 10:619–624.
55. Olesik JW, Thaxton KK, Kinzer JA, Grunwald EJ. Paper T8, Winter Conference on Plasma Spectrochemistry, Scottsdale, AZ, 1998.

56. Miller-Ihli J. Paper T10. Winter Conference on Plasma Spectrochemistry, Scottsdale, AZ, 1998.
57. Szpunar J, Chassaigne H, Donard OFX, Bettmer J, Lobinski R. Applications of ICP-MS. Holland G, Tanner S, eds. Cambridge, England: Royal Society of Chemistry, 1997.
58. Al-Rashdan A, Heitkemper D, Caruso JA. J Chromatogr Sci 1996; 29:98.
59. Ding H, Olson LK, Caruso JA. Spectrochim Acta Part B 1996; 51:1801.
60. Shum SCK, Nedderden R, Houk RS. Analyst 1992; 117:577.
61. McLean JA, Zhang H, Montasser A. Anal Chem 1998; 70:1012–1020.

18

ICP-MS Applications

Today, there are over 5000 ICP-MS installations worldwide, performing a wide variety of applications, from routine, high-throughput multielement analysis to trace element speciation studies using high-performance liquid chromatography. Every year, as more and more of the trace element user community realizes the benefits of ICP-MS, the list of applications gets bigger and bigger. In this chapter, we will take a look at the major market segments addressed by ICP-MS, such as environmental, biomedical, geochemical, semiconductor, and nuclear and give detailed examples of the most common types of applications being carried out.

As a result of the widespread use and acceptability of ICP-MS, the cost of commercial instrumentation has dramatically fallen over the past 20 years. When the technique was first introduced, $250,000 was a fairly typical amount to spend, whereas today, you can purchase a system for less than $150,000. Although it can cost a great deal to invest in magnetic sector technology or a quadrupole instrument fitted with a collision/reaction cell, most laboratories that are looking to invest in the technique should be able to justify the purchase of an instrument based without price being a major concern. One of the benefits of this kind of price erosion is that slowly but surely, the AA and ICP-OES user community are being attracted to ICP-MS, and, as a result, the technique is being used in more and more diverse application areas. Figure 18.1 shows a percentage breakdown of the major market segments being addressed by ICP-MS on a worldwide basis. Two points should be emphasized here. First, these data can be significantly different on a geographical basis because of factors like a country's commitment (or lack of it) to environmental concerns or the size of a region's electronics or nuclear industry, for example. Secondly, many laboratories carry out more than one type of application and, as a result, can be represented in more than one market segment. For these reasons, these data should only be considered an approximation for comparison purposes.

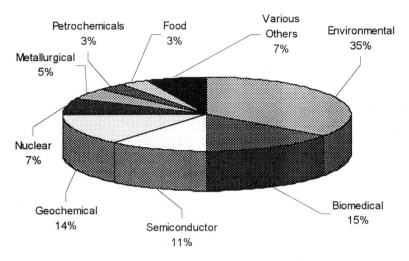

FIGURE 18.1   Breakdown of major market segments addressed by ICP-MS.

Let us now take a look at each of these market segments in greater detail. The intent in this chapter is to present a broad cross section of application work being carried out in each market segment They represent typical analytical problems, but in no way are meant to be a comprehensive list of all application work being addressed by ICP-MS. Where possible, I have suggested further reading with additional published literature references.

ENVIRONMENTAL

As can be seen by the pie chart, environmental applications represent the largest market segment for ICP-MS. In fact, about a third of all applications being carried out are environmental in nature. The most common types of environmental samples being analyzed today for trace element determinations include drinking waters, groundwaters, wastewaters, river waters, estuarine waters, seawaters, solid waste, soils, sludges, sediments, and airborne particulates. There is no question that the enormous growth in environmental applications, especially in North America, is based on legislature driven by the U.S. Environmental Protection Agency (EPA) (http://www.epa.gov). Environmental users are generally not pushing the extreme detection capability of ICP-MS. This can be seen in Table 18.1, which compares the National Primary Drinking Water Regulations (NPDWR)

TABLE 18.1 Comparison of ICP-MS Detection Limits with NPDWR Maximum Contaminant Levels for the 12 Primary Trace Metal Contaminants in Drinking Water

Element	NPDWR MCL (μg/L)	Typical ICP-MS DLs (μg/L)
As	10	0.05
Ba	2000	0.01
Be	4	0.01
Cd	5	0.02
Cr	100	0.05
Hg	2	0.01
Pb	15	0.005
Ni	100	0.005
Cu	1300	0.005
Sb	6	0.002
Se	50	0.2
Tl	2	0.001

Source: Ref. 1.

maximum contaminant levels (MCL) with typical ICP-MS detection limits for the 12 primary contaminants in drinking water.

These levels are covered by EPA Method 200.8 (1), which is approved for all 12 primary contaminants in drinking water (shown in Table 18.1) and most of the secondary ones including Al, Mn, Ag, and Zn. It should be noted that in January of 2001, the MCL goal for arsenic (As) in drinking water was set at zero (2). This was a health-based initiative and was not actually enforceable. However, in February of 2002, an enforceable MCL of 10 ppb was applied to community water systems and noncommunity water systems, which are not presently subject to arsenic standards. In addition, the EPA Office of Water (http://www.epa.gov/ow) has stated that all water systems nationwide must be fully compliant by January 2006. This extremely low level means that only ICP-MS or GFAA (under Method 200.9) methods can be used to determine arsenic because ICP-OES methodology (inc. Method 200.7) cannot meet the required limits of quantitation.

In addition to drinking waters, Method 200.8 can also be used for trace elements in wastewater—under the National Pollutant Discharge Elimination System (NPDES). It has had general approval since 1995, but full acceptance varies on a regional basis, which means that each lab must apply for an Alternate Test Procedure (ATP) to their local EPA Quality Assurance

Officer. In addition, since January 2000, Method 200.8 can also be used under NPDES rules for the analysis of wastewaters from industrial incinerators (3). Other Office of Water ICP-MS-related methodology include:

> Method 1638, which is a variation of Method 200.8, for the determination of trace elements in ambient waters (4)
>
> Method 1640 for the determination of trace metals in ambient waters by on-line chelation and preconcentration (5)
>
> Method 1669 for the sampling of ambient water for the determination of trace metals at EPA water quality criteria levels (6)

In addition to the Office of Water, the Office of Solid Waste and Emergency Response (OSWER), which conducts overseas land disposal of solid waste, underground storage tanks, hazardous waste, and Superfund sites, also has approved a number of ICP-MS-related methods. They include:

> SW-846 Method 6020, covered by the Resource Conservation and Recovery Act (RCRA) program for monitoring 15 trace metal contaminants (Al, Sb, As, Ba, Be, Cd, Cr, Co, Cu, Pb, Mn, Ni, Ag, Tl, and Zn) in hazardous waste, solid waste, industrial waste, soils, sludges, sediments, and groundwaters (7).
>
> Update IVA (Federal Register Vol. 63, p. 25430) contains all methods which are being considered for inclusion in SW-846.
>
> Method 6020A (similar to Method 6020, but more performance-based) was proposed in an update in 1998 to include additional 8 elements— Ca, Fe, Mg, Hg, K, Se, Na, and V (8).
>
> Latest EPA Statement of Work for Inorganic Analysis actually incorporates ICP-MS instead of GFAA (9).
>
> Method 6080 which covers the determination of elemental species by isotope dilution mass spectrometry.
>
> Method 6020-CLP-M under the Contract Laboratory Program which is available as a Special Analytical Services (SAS) method.

It should be emphasized that the EPA is continually looking to update their methods based on new technology and input from the trace metal user community, so although this information represents the state of the methodology at the time of writing this book, you should always check on their current status if you have particular questions or concerns. However, when all these methods are added to all the other ICP-MS-based methodology recommended by other standards organizations including Department of Energy (DOE), American Water Works Association (AWWA), American Standard Test Methods (ASTM), and the huge growth in speciation studies using chromatography separation devices coupled to ICP-MS (10), it makes the technique a very attractive option for environmental labs. It means they can now determine the vast majority of the environmentally significant elements/species by one technique. This capability is very attractive because it means they can typically analyze 5–10× more samples per day, for a full suite of elements, compared to other approaches that use a combination of FAA, GFAA, CVAA (cold vapor for Hg), and ICP-OES. This productivity improvement is exemplified in Table 18.2, which compares the productivity of a drinking water analysis for 12 primary contaminants using three different analytical scenarios (11).

The first scenario using GFAA, FAA, and CVAA is typical of smaller laboratories that do not have ICP-OES capability. The second scenario using GFAA, CVAA, and ICP-OES is typical of many larger environmental laboratories, while the third scenario uses ICP-MS for the full suite of analytes, including mercury. It can be seen by the number of samples analyzed in an 8-hr shift that the productivity with ICP-MS is significantly higher than with the other multitechnique approaches. This productivity enhancement clearly translates into a reduction in the overall cost of analysis resulting in a much faster instrument payback period.

These EPA-driven methods represent the bulk of the routine environmental analysis being carried by ICP-MS today. However, there are many other types of samples being analyzed, which represent a much smaller but significant contribution to the environmental application segment. For example, in order to better understand industrial-based airborne pollution covered by the Clean Air Act, air quality is often monitored using air filtering systems. These typically consist of small pumps (either static or personal) where the air is sucked through a special filter for extended periods of time. The filter paper is then removed, dissolved in a dilute acid, and analyzed by an appropriate

TABLE 18.2 Productivity Comparison Between ICP-MS and Other Multitechnique Approaches for the Determination of 12 Primary Contaminants in Drinking Water

Technique	Scenario 1: GFAA/FAA/CVAA	Scenario 2: GFAA/ICP-OES/CVAA	Scenario 3: ICP-MS
GFAA+FLAA	440 min	160 min	—
ICP-OES	—	70 min	—
Hg Prep	120 min	120 min	—
Hg Analysis	40 min	40 min	—
ICP-MS	—	—	74 min
Total time	600 min	390 min	74 min
Sample/8 hr	16	25	130

Source: Ref. 11.

technique. Because trace metal concentration levels are sometimes extremely low, ICP-MS has proved itself to be a very useful tool to analyze these airborne particulate samples and help pinpoint sources of industrial pollution.

Other important work involves the analysis and the classification of environmental-based certified reference materials produced by standards organization like National Institute of Standards and Technology (NIST) and National Research Council of Canada (NRC). Some of these standards include drinking waters, river waters, open ocean seawaters, coastal seawaters, estuarine waters/sediments, freeze-dried dogfish/muscle tissue, spinach/orchard leaves, and many more. These reference materials are often analyzed using isotope dilution methods (refer to Chapter 13, "Methods of Quantitation") because traditional external calibration typically does not offer high-enough accuracy (12).

However, it should be emphasized that probably 90% of all routine environmental labs are using basic quadrupole ICP-MS instrumentation. That is not to say other types of mass analyzers are not suitable for environmental analysis, but when detection limit requirements, sample throughput demands, operator skill level, and financial considerations are taken into account, quadrupole technology is the logical choice. In fact, vendors are now beginning to offer turnkey systems containing all aspects of EPA methodology, including analyte masses, internal standards, integration times, QC protocol, etc. These methods are designed specifically for environmental users because the majority of instruments are being operated by technicians, with limited experience in ICP-MS. For more information on the analysis of environmental samples by ICP-MS, Refs. 13–16 should be helpful.

BIOMEDICAL

The second-largest market segment is biomedical. Compared to other markets like environmental and geochemical, the biomedical community was relatively late in realizing the benefits of ICP-MS as a routine tool. Although early biomedical researchers showed the capabilities of ICP-MS (17,18), it was not until the early 1990s that it was first used as a technique for routine nutritional and toxicity studies (19). Since then, it has probably become the fastest-growing market segment for ICP-MS because it provides a fast, cost-effective way to carry out trace element studies in critically important areas of biomedical research such as toxicology, pathology, nutrition, forensic science, occupational hygiene, and environmental contamination. Some of the many kinds of biomedical analyses being carried out by ICP-MS include

> Determination of toxic elements, like As, Cd, and Pb in blood—as an indication of whether a person could be exposed to some kind of

contamination in their home or from industrial-based pollution of the environment (20).

It is important to know the levels of nutritional elements like Fe, Cu, and Zn in human serum to understand how they are absorbed into the bloodstream (21).

Monitoring Al in patients who are undergoing kidney dialysis treatment (22).

The determination of trace elements in bones and teeth as an indicator to heavy metal exposure (23).

The multielement analysis of hair samples can indicate whether a person is lacking in essential vitamins and nutrients (24).

As you would expect, the analysis of clinical-type samples is not that straightforward because of the complex nature of blood, urine, serum, body tissue samples, etc. Unlike environmental samples, which often require just simple acidification or maybe an acid digestion, the matrix components of biomedical samples can pose some unique problems for ICP-MS in the areas of sample preparation, interference correction, calibration, and long-term stability (25). Let us take a look at these in greater detail.

Sample Preparation

Ideally, the sample preparation methods must be simple, straightforward, and be able to be carried out in a routine manner. The more complex is the sample preparation, the greater the chance of contamination, which ultimately affects accuracy and spike recoveries. The preferred method of sample preparation is by simple dilution with a suitable diluent like dilute nitric acid for urine or 5–10% tetra methyl ammonium hydroxide (TMAH) for blood. However, this is not always possible with all types of biological materials. In these cases, a digestion with concentrated HNO_3 acid followed by filtration or centrifuging may be required to leach all elements into solution. If this type of sample preparation is required, microwave digestion apparatus has simplified the digestion of difficult samples and is usually the preferred approach over conventional hot plate acid digestion.

Interference Corrections

During method development, special attention must be given to correct for matrix and spectral interferences. Matrix suppression and sample transport interferences are compensated very well by the selection of suitable internal standards, which are matched to the ionization properties of the analyte elements. This is a routine and well-understood method for compensating for matrix-related interferences. However, a more serious problem in the analysis of clinical samples is that analytes of interest can be affected by isobaric, polyatomic, and molecular spectral interferences resulting from plasma and matrix species. Table 18.3 shows some common interferences seen in clinical samples.

To get around this problem using a basic quadrupole system, either another isotope of the element of interest has to be monitored or an elemental correction equation needs to be applied. This is common methodology used to analyze clinical samples. However, if the trace metal levels in the sample are extremely low, or sample preparation necessitates the use of an acid/solvent that contains one of the interfering ions (e.g., Cl^+ or N^+), this approach struggles. For that reason, ultratrace levels in some clinical samples either require the use of a high-resolution magnetic sector instrument to resolve the interference away or collision/reaction cell technology to stop the formation of the interference using ion molecule chemistry.

Calibration

Because of the differences in the matrix components of samples like urine, blood, or serum, simple external calibration can often produce erroneous results. For that reason, it is common to use other calibration methods like standard additions or additions calibration to achieve accurate data. These methods have been described in detail in Chapter 13, "Methods of Quantitation," but they are required because of the matrix suppression effects caused by large variations in patients' biological fluid samples. The sample preparation method used will often dictate the type of calibration curve to use, but all three methods are all absolutely necessary to achieve good accurate data when analyzing clinical samples by ICP-MS.

TABLE 18.3 Some Common Spectral Interferences Seen in Clinical Matrices

Element	Interference
$^{24}Mg^+$	$^{12}C^{12}C^+$
$^{27}Al^+$	$^{13}C^{14}N^+$
$^{51}V^+$	$^{16}O^{35}Cl^+$
^{52}Cr	$^{40}Ar^{12}C^+, {}^{16}O^{35}ClH^+$
$^{58}Ni^+$	$^{58}Fe^+, {}^{42}Ca^{16}O^+$
$^{63}Cu^+$	$^{40}Ar^{23}Na^+$
$^{75}As^+$	$^{40}Ar^{35}Cl^+$
$^{80}Se^+$	$^{40}Ar^{2+}$

Stability

Today, ICP-MS has proven rugged enough to be used routinely in high-throughput clinical laboratories. However, complex blood, urine, serum, and digested body tissue matrices can affect signal stability, resulting in the need for frequent recalibration. One of the major problems is that matrix components (salts, carbon, proteins, etc.) can deposit either on the tip of the plasma torch sample injector or on the orifice of the sampler and/or skimmer cone, which, over time, can eventually lead to blockage and signal instability. Another negative impact of clinical matrices on an ICP mass spectrometer is that material can deposit itself on the ion optics system, leading to instability and the likelihood of reoptimizing the lens voltages. Although some instrument designs will be affected less than others, it is well accepted that routine maintenance (including regular cleaning and/or replacing parts) is absolutely essential to keep up with the harsh demands of running clinical samples, especially if the instrument is being used on a routine basis.

Although levels of interest are generally lower than those required by the environmental ICP-MS community, the biomedical market segment is interested in a similar suite of elements and also has similar sample throughput and productivity demands. This has been driven by a growing demand to bring down the cost of analysis to lessen the financial burden on hospitals and health authorities. All these factors have contributed to the overwhelming acceptance of ICP-MS for the trace element analysis of biomedical samples, in preference to slower, less productive techniques like GFAA.

It is also worth mentioning that understanding the effects of different elemental forms and species on human health and its impact on the environment has sparked an enormous growth in speciation studies using ICP-MS and chromatography separation devices like liquid chromatography (HPLC) (26,27,28), size exclusion (SEC) (29), supercritical fluid extraction (SFEC) (30), and capillary zone electrophoresis (CZE) (31). This has been described in greater detail in Chapter 17, "Alternate Sampling Accessories."

GEOCHEMICAL

Geochemists were some of the first researchers to realize the enormous benefits of ICP-MS for the determination of trace elements in digested rock samples (32). Up until then, they had been using a number of different techniques including neutron activation analysis (NAA), thermal ionization mass spectrometry (TIMS), plasma emission (ICP-OES), x-ray techniques, and GFAA. Unfortunately, they all had certain limitations, which meant that no one technique was suitable for all types of geochemical samples. For example, NAA was very sensitive, but when combined with radiochemical separation techniques for the determination of rare earth elements, it was extremely slow and expensive to run (33). TIMS was the technique of choice for carrying out isotope ratio studies because it offered excellent precision, but, unfortunately, was painfully slow (34). Plasma emission was very fast and excellent for multielement analysis, but it was not very sensitive. In addition, because the technique suffered from spectral interferences, ion-exchange techniques often had to be used to separate the analyte elements from the rest of the matrix components (35). X-ray techniques like XRF (fluorescence) were rapid, but were generally not suited for ultralow levels and also struggled with some of the lighter mass elements (36). While GFAA had good sensitivity, it was predominantly a single element technique and was therefore very slow (37). It was also not suitable for low levels of refractory or rare earth elements because the low atomization temperature of the electrothermal heating device ($<3000\,°C$) did not produce sufficiently high numbers of ground-state atoms. Although all these techniques are still used to some degree, all these factors led to the very rapid acceptance of ICP-MS by the geochemical user community.

Geochemists represent some of the most demanding users of ICP-MS. Invariably, they are looking for ultratrace levels in the presence of large concentrations of major elemental components in digested rock samples, like Ca, Mg, Si, Al, and Fe. This alone presents difficulties for the sample introduction and interface region because of the potential for signal drift caused by the geological material depositing itself on the cones and ion lens system. In addition, if there are large concentrations of high-mass elements like Tl, Pb, or U present in the sample, they can cause severe space-charge matrix suppression on the analyte masses. Another potential problem is that major and trace components in the sample can combine with argon-, solvent-, and acid-based species to produce quite severe polyatomic, isobaric, doubly charged, and oxide-based spectral interferences. When this is combined with the extremely demanding sample preparation methods using highly corrosive materials like concentrated aqua regia (HCl/HNO3), hydrofluoric acid (HF), and/or fusion mixtures to dissolve the samples, it makes the geological matrices some of the most difficult to analyze by ICP-MS. Let us now highlight some of these problem areas by taking a look at some typical geochemical applications being carried out by ICP-MS.

Determination of Rare Earth Elements

The determination of rare earth elements was one of the very first applications that attracted geochemists to ICP-MS mainly because of the lengthy sample preparation and analysis times involved with previously used techniques like ICP-OES and NAA (38). However, although ICP-MS offered significant benefits over these techniques, it was not without its problems because of the potential of spectral interferences from other rare earth elements in rocks or natural water samples. For that reason, instrument parameters have to be optimized, depending on the rare earth elements being determined and the kinds of interferents present in the sample. For example, plasma power and nebulizer gas flows must be adjusted to minimize the formation of oxide species. This is necessary because an oxide or hydroxide species of one rare earth element can spectrally interfere with another rare earth element at 16 or 17 amu higher. The problem can be alleviated by using a sample desolvation device like a chilled spray chamber to reduce oxide formation but, unfortunately, cannot be completely eliminated. For that reason, to get the best detection capability for rare earth elements in geological matrices, instrument sensitivity must often be sacrificed for low oxide performance and even then, mathematical correction equations might need to be applied. One of the many examples of this type of interference is the contribution of praseodymium oxide ($^{141}Pr^{16}O^+$) at 157 amu on the signal of $^{157}Gd^+$, one of the major isotopes of gadolinium. Other examples of rare earth elements that readily form oxides/hydroxides, and the elements they interfere with, are shown in Table 18.4 (38).

It is also worth pointing out that in addition to the formation of oxide species, some rare earth elements can generate high levels of doubly charged ions (ions with two positive charges as opposed to one). This is not so much of a problem with the determination of other rare earth elements, but more their spectral impact on other lower mass analytes. Examples of rare earth elements

TABLE 18.4 Examples of Rare Earth Elements that Readily Form Oxide and Hydroxide Species in ICP-MS

Rare earth oxide/hydroxide	Interferes with
$^{135}Ba^{16}O^+$	$^{151}Eu^+$
$^{136}Ba^{16}O^+$, $^{136}Ce^{16}O^+$	$^{152}Sm^+$
$^{141}Pr^{16}O^+$, $^{140}Ce^{16}OH^+$	$^{157}Gd^+$
$^{143}Nd^{16}O^+$, $^{142}Ce^{16}OH^+$	$^{159}Tb^+$
$^{146}Nd^{16}OH^+$, $^{147}Sm^{16}O^+$	$^{163}Dy^+$
$^{149}Sm^{16}O^+$	$^{165}Ho^+$
$^{152}Sm^{16}O^+$	$^{168}Er^+$
$^{153}Eu^{16}O^+$, $^{152}Sm^{16}OH^+$	$^{169}Tm^+$
$^{158}Gd^{16}O^+$	$^{174}Yb^+$
$^{158}Gd^{16}OH^+$, $^{159}Tb^{16}O^+$	$^{175}Lu^+$

Source: Ref. 38.

that easily form doubly charged species include barium, cerium, samarium, and europium as shown in Table 18.5. If these elements are present in high-enough concentrations, certain isotopes can interfere with analytes at one-half of their mass. Parameter optimization can help, but even more important is to minimize the effects of high plasma potential (a secondary discharge at the interface is known to increase doubly charged species) with well-grounded RF coil (39). However, with certain geological matrices, no matter what precautions are taken, doubly charged species are unavoidable depending on the analytes of interest.

Analysis of Digested Rock Samples Using Flow Injection

The benefits of flow injection (FI) techniques for ICP-MS have been described in detail in Chapter 17, "Alternate Sampling Accessories." The main advantage of FI for the analysis of geological samples is the ability to aspirate high concentrations of dissolved solids into the mass spectrometer. With continuous nebulization, it is well accepted that to maintain good stability, the total dissolved solids (TDS) in the sample should not exceed 0.2% w/v, which can be a severe limitation if analyte concentrations are extremely low. However, using the microsampling capability of FI, where small volumes (typically <500 μL) of the sample are transported into the ICP-MS in a continuous flow of carrier liquid, much larger levels of dissolved solids can be tolerated. In fact, it is fairly common to put in excess of 1% w/v dissolved into the ICP-MS system using this approach and still maintain good accuracy and precision for geological matrices. This is exemplified in Table 18.6, which shows the determination of a group of elements in a United States Geological Survey (USGS) standard reference rock (andesite)—AGV-1, using USGS SRM BEN (basalt) for calibration. Both sample and calibration standard were dissolved using a lithium tetraborate ($Li_2B_4O_7$) fusion mixture, which, including weight

TABLE 18.5 Examples of Rare Earth Elements that Readily Form Doubly Charged Species and the Analyte Masses They Interfere With

Doubly charged species	Interferes with
$^{138}Ba^{2+}$	$^{69}Ga^+$
$^{140}Ce^{2+}$	$^{70}Ge^+$, $^{70}Zn^+$
$^{151}Eu^{2+}$	$^{75}As^+$
$^{152}Sm^{2+}$	$^{76}Ge^+$, $^{76}Se^+$

TABLE 18.6 Determination of a Group of Elements in a USGS Standard Reference Rock (Andesite)—AGV-1 Using Flow Injection Microsampling (TDS in solution was 1.2% w/v)

Element	USGS AGV-1 reference value (mg/kg)	Measured value by FI-ICP-MS (mg/kg)	Precision (% RSD)
Ba	1226.0	1204.0	1.0
Be	2.1	2.1	2.1
Ce	67.0	70.5	0.6
Co	15.3	14.4	1.2
Cs	1.3	1.7	1.8
Cu	60.0	52.3	0.9
Eu	1.6	1.5	4.6
Ga	20.0	20.2	0.4
La	38.0	33.7	1.4
Lu	0.3	0.2	4.1
Mo	2.7	2.0	2.3
Sr	662.0	628.3	0.6
Yb	1.7	1.4	0.6
Zn	88.0	111.3	0.8
W	0.6	0.8	2.5
V	121.0	121	0.0
U	1.9	1.8	6.6

Source: Ref. 40.

of sample, represented 1.2% w/v total dissolved solids—6× more material than is typically aspirated into an ICP mass spectrometer (40).

Geochemical Prospecting

Exploring for deposits of the platinum group elements (PGE), commonly known as precious metals, is typically carried out by sampling large areas to establish a concentration contour map. The occurrence, distribution, and concentration of these precious metal deposits are then used to ascertain whether mining is economically feasible in that area. Analytical methodology developed for the determination of precious metals in geological samples must therefore have sufficient sensitivity to quantify individual PGEs at the ng/g (ppb) level, but also be fast enough in order to cost-effectively handle such a large number of samples (41).

Determination of precious metals in geological samples is generally a three-step process. The first step involves preparing a representative sample (which can be challenging in itself) and then isolating the analyte from the ore matrix using established methods like the fire assay technique. This typically involves fusion with a flux, producing a lead or nickel sulfide button, which is then ground into a powder. The second step separates the precious metals from the rest of the matrix by a process called cupellation. This process involves heating the powdered sample in a cupel made of bone ash (phosphate of lime), where the matrix components are oxidized into the porous cupel, leaving the precious metals separated out from the rest of the sample (42). The final step involves dissolution of the precious metals with a suitable acid and measurement of the analyte concentrations by some sort of instrumental technique. There are slight variations to the fire assay procedure (based on how the PGE is extracted from the lead/nickel sulfide button), which is often dictated by the type of sample being collected and the elemental requirements of the analysis. The result is that a number of different trace element techniques have been used for this type of analysis, including GFAA, ICP-OES, and NAA. All three approaches have been used to quantify PGEs with good accuracy in fire assay samples, but as mentioned earlier, ICP-OES will struggle with low concentrations, and in the case of GFAA and NAA, sample throughput is severely restricted because of its slow speed of analysis. All these factors have contributed to the rapid acceptance of ICP-MS for the determination of PGEs by fire assay (43) and other sample preparation methods (44). This is emphasized in Figure 18.2, which shows the superior detection capability of ICP-MS over both GFAA and ICP-OES for the PGEs.

The ultralow detection capability of ICP-MS, combined with its rapid speed of analysis, high sample throughput, and excellent accuracy and precision, makes it ideally suited for this type of work. In fact, in countries like Australia and Canada that have large mineral deposits, large commercial labs

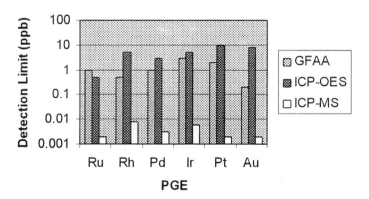

FIGURE 18.2   Detection capability improvement of ICP-MS over GFAA and ICP-OES for the platinum group elements.

have sprung-up that use ICP-MS on a 24-hr basis to support their country's extensive mining exploration business (41).

Isotope Ratio Studies

The study of isotope ratios is extremely important to geochemists and environmentalists, both as a means of approximating the age of rock formations (geochronology) (45) and tracing the source of metallic pollutants on the environment and ecosystems (46). However, one of the main requirements for this kind of analysis is the ability of the method to generate extremely good precision data. For this reason, the most widely used instrumental approach has involved the use of thermal ionization mass spectrometry (TIMS) (47). Unfortunately, although TIMS is capable of producing isotope ratio precision down to 0.005% RSD, the analytes have to be isolated from the matrix, making the sample pretreatment cumbersome and time-consuming. In addition, the sample solutions obtained have to be preconcentrated and loaded onto a filament, which are then mounted onto a sample turret and subsequently inserted into the vacuum-pumped chamber of the TIMS instrument.

These sample throughput limitations led geochemical researchers to investigate ICP-MS as a possible solution to their problem. Unfortunately, it soon became clear that although quadrupole ICP-MS demonstrated excellent throughput, the best isotope precision it could offer on a routine basis was 0.2–0.5% RSD. It was not until the commercialization of double-focusing magnetic sector ICP-MS technology in the early 1990s did geochemists realize that they had an analytical tool that could perhaps compete with TIMS for carrying out isotope ratio studies (48). The extremely high sensitivity, low background, fast scanning, and flat-topped peaks of this technique consistently demonstrated precision data in the order of 0.05–0.10% RSD as can be seen in Table 18.7, which shows $^{206}Pb^+/^{207}Pb^+$ isotope ratio precision data,

TABLE 18.7 Typical $^{206}Pb^+/^{207}Pb^+$ Isotope Precision Data for a Single Collector, Double Focusing Magnetic Sector ICP-MS, Compared to Its Statistical Counting Limits

Measurement set	Experimental RSD for $^{206}Pb^+/^{207}Pb^+$ ($n = 10$)	Theoretical RSD (based on counting statistics)
1	0.11	0.062
2	0.044	0.062
3	0.12	0.065
4	0.063	0.053

taken from a paper by Vanhaecke et al. (48). It should be noted that the lead concentration data were varied slightly in each measurement set in order to produce a peak height of ~200,000 cps, and it can be clearly seen that the experimental data for each set are approaching its statistical counting limits.

Although this kind of data was much better than quadrupole ICP-MS, it still was not as good as TIMS. For this reason, the geological community wanted even better isotope ratios by ICP-MS, and as a result, instrument manufacturers eventually answered their demands with the development and the commercialization of multicollector, magnetic sector ICP-MS systems. This design, which utilized multiple detectors instead of just one, allowed for the simultaneous measurement of each mass, offering the capability to generate isotope ratios equivalent to TIMS (49).

Such are the extreme demands of the geochemical application sector, that researchers are looking for techniques and sampling accessories that offer a high level of performance and flexibility. For that reason, the high-resolution, high-sensitivity, and excellent precision capability of magnetic sector systems make them ideally suited to this kind of work. In addition, collision/reaction cell technology is becoming more popular with geochemists because of its ability to chemically resolve away many of the spectral interferences using ion molecule chemistry. In fact, Vanhaecke et al. (50) have recently shown that $^{87}Rb^+$ can be "chemically resolved" from $^{87}Sr^+$ using a mixture of methyl fluoride (CH_3F) and neon (Ne) gas in a dynamic reaction cell, so that $^{87}Sr^+/^{86}Sr^+$ isotope ratios can be measured with good-enough precision for geochronological dating studies. The attraction of this technology over a high-resolution approach is that it would take a resolving power of 290,000 to separate $^{87}Rb^+$ from $^{87}Sr^+$ using magnetic sector technology. In addition, if there is significant amount of krypton in the argon supply (which is fairly common), it would require a resolving power of 66,000 to separate $^{86}Kr^+$ from $^{86}Sr^+$. Unfortunately, this is way beyond the capabilities of commercial magnetic sector instruments, which typically offer resolving power up to 10,000.

Laser Ablation

Laser ablation as a sampling tool and its applicability for use with ICP-MS has been described in great detail in Chapter 17, "Alternate Sampling Accessories." There is no question that after many years of being considered a "novel and interesting" technique, it has now been refined to become an extremely useful sampling technique for many types of materials (51). However, it was primarily geochemists and mineralogists who drove the development of laser ablation for ICP-MS because of their desire for ultratrace analysis of optically challenging materials, such as calcite, quartz, glass, and fluorite, combined with the capability to characterize small spots and micro-inclusions on the surface of the sample. For that reason, most of the fundamental studies into the ablation process have been based on the analytical demands of the geochemical community (52).

However, there is still much debate as to the optimum design to use for the many diverse types of geochemical samples (53). Based on literature in the public domain, the general consensus today is that 266 nm Nd:YAG technology is extremely good for bulk analysis and for some of the less-challenging microinclusion work. On the other hand, the shorter wavelength 213 nm Nd:YAG technology couples much better with UV transparent material like silicates, fluorites, and calcites and is also better suited for the study of minute fluid inclusions because of its more controlled ablation process. On the other hand, 193-nm ArF excimer laser offers the most precise ablation character-istics of all three designs and excels with the most optically challenging ma-terials (54). It is also the best tool to use for precise and accurate depth measurement studies. The main disadvantage of the ArF excimer design is that because of its optical complexity and the requirement to use a toxic gas, it requires a more skilled person to operate and maintain it. It is also the most expensive of all three designs to purchase.

So, although the development of laser ablation for ICP-MS has gone in many different directions, it is now generally accepted that shorter wave-lengths are considered more suited for geological matrices, especially for the analysis of small spots and inclusions in UV transparent materials. However, it is still a very active area of research, which sees new developments and refinements on a regular basis. Although arguments can be made for the benefits of one specific design over another, it is not that straightforward, especially when the capability of the ICP mass spectrometer is taken into consideration. For example, when it is being applied to the analysis of micro-inclusions, it is absolutely critical that the ICP-MS system is capable of very high sensitivity because you may only be able to fire one laser pulse to ablate the area of interest. In addition, when analyzing a fast transient ($\sim$10 sec) of fine particles generated by a single laser pulse, it is very important that the scanning and settling times of the mass analyzer are kept to an absolute minimum. For these reasons, it is fairly common to see double focusing mag-netic sector ICP-MS technology used with laser ablation systems because of its extremely high sensitivity (55). Also, TOF is beginning to show its benefits for laser ablation work because of its ability to simultaneously sample the ion beam and capture the maximum amount of data in the limited duration of the short transient peak (56).

So the optimum combination of laser ablation system and ICP mass spectrometer can often be sample- and application-specific. There is no question that in the hands of a good operator, most laser ablation systems should work well with any commercial ICP-MS systems and be capable of generating data of the highest quality on complex rock samples, small inclusions in rocks, or carrying out elemental surface mapping studies. It is not the intent of this book to show bias towards any design, but just to emphasize that if there is a need for this kind of solid-sampling capability, each integrated system should be evaluated on a sample-to-sample basis. In addition, good literature references should be read in order to get a better understanding of what features are important for geochemical analysis (57).

SEMICONDUCTOR

The semiconductor industry is probably the most demanding user of ICP-MS, with regard to its detection capability requirements. Consumer demand for smaller electronic devices and more compact integrated circuits has resulted in the need for ultratrace metal contamination levels on the surface of silicon wafers and also in the high-purity chemicals and gases used in various stages of the semiconductor manufacturing process. In order to reduce costs and increase yield, chip manufacturers are making larger diameter wafers with even narrower line widths. This trend, which is being driven by initiatives like the International Technology Roadmap for Semiconductors (ITRS) (58), is setting the course for the next generation of semiconductor devices and has resulted in lower trace element contamination levels in all semiconductor-related materials. Whereas 10 years ago, the Semiconductor Equipment and Materials International (SEMI) organization deemed that 1 ppb purity levels were adequate for many of the process chemicals; today, 100 ppt is typical—and for some of the more critical materials, 10 ppt guideline levels are currently being proposed (59).

The SEMI Book of Semiconductor Standards (BOSS) has approved ICP-MS for the determination of trace metals in a number of chemicals at the 10 ppb (Grade 3) and 100 ppt (Grade 4) levels and is looking into the feasibility of approving the technique for some chemicals at the 10 ppt (Grade 5) level. Table 18.8 shows typical specification levels for all the semiconductor-significant elements. Some element specifications are different for different chemicals, but this table represents a good approximation of the trend for comparison purposes (59).

However, the BOSS states that GFAA can also be used if ICP-MS does not have the required detection limit. The inherent problem lies in the fact that many of these corrosive chemicals have to be diluted 10× or even 100× to aspirate them into the ICP mass spectrometer, which obviously degrades detection capability in the original chemical. In addition, for a SEMI guideline (proposed specification) to be converted into a published standard, analytical data must be generated, which shows spike recovery data between

TABLE 18.8 Trend in Specification Levels for the
Semiconductor-Significant Elements (ppb)

Element	Grade 3	Grade 4	Grade 5
Aluminum	1000	100	10
Antimony	1000	100	10
Arsenic	1000	100	10
Barium	1000	100	10
Boron	1000	100	10
Cadmium	1000	100	10
Calcium	1000	100	10
Chromium	1000	100	10
Cobalt	1000		
Copper	1000	100	10
Gold	10000		
Iron	1000	100	10
Lead	1000	100	10
Lithium	1000	100	10
Magnesium	1000	100	10
Manganese	1000	100	10
Molybdenum	1000		
Nickel	1000	100	10
Potassium	1000	100	10
Silver	1000		
Sodium	1000	100	10
Strontium	1000		
Tin	1000	100	10
Titanium	1000	100	10
Vanadium	1000	100	10
Zinc	1000	100	10

Source: Ref. 59.

75% and 125% at 50% of the proposed specification level. Only when this happens will it be published in the BOSS as an official specification or standard (59). So if SEMI proposes that a chemical needs to be at a 100-ppt purity level (Grade 4), the technique has to prove that it can meet spike recovery data at the 50-ppt level for the full suite of elements. However, if a 10× dilution of the chemical has to be made, which is typical for semiconductor chemicals, spike recovery data at the 5-ppt level has to be shown. Clearly, traditional ICP-MS will struggle because for many critical elements, this is only 5–10× above the detection limit. The problem is even more serious when the technique has to demonstrate spike recovery data for Grade 5 (10 ppt) purity levels.

Over the years, instrument manufacturers have adapted and refined the technique to improve detection capability for many elements. A prime example of this is the development of the cool or cold plasma technique, which has been described in greater detail in Chapter 14, "Review of Interferences." Up until its commercialization in the early–mid-1990s, detection limits for the notoriously difficult ICP-MS elements like Fe, Ca, and K were in the order of parts per billion levels because of high spectral background levels from argon-derived polyatomic ions. Using cool plasma conditions, which involved a decrease in RF power, an increase in nebulizer gas flow, and sometimes a change in the sampling depth, these argon-based polyatomic interferences could be reduced to such an extent that detection limits in the order of 1–5 ppt could be achieved (60). This innovation meant that Grade 4 purity levels for chemicals like 30% hydrogen peroxide and 5% hydrofluoric acid (which required no dilution step) could be achieved without the need for GFAA. This represented a major saving in time and money because ICP-MS alone could determine the full suite of elements. However, even cool plasma technology struggled with Grade 4 levels of Fe, K, and Ca if a sample dilution was required. In addition, the lower ionization temperature of a cool plasma degraded detection limits for some elements, which necessitated the use of both normal and cool plasma conditions for a multielement run. The lower temperature was even more of a problem when analyzing matrices like concentrated acids, caustic solutions, and complex organic solvents because of severe matrix suppression on the analyte signal.

So it was clear that to meet the proposed Grade 5 purity levels (10 ppt in chemical), traditional quadrupole instruments, even using cool plasma conditions, were not going to have the detection capability for the full suite of semiconductor elements. This limitation in quadrupole technology opened the door to other approaches like high-resolution, magnetic sector technology, which offered both high sensitivity and the ability to resolve the polyatomic interferences away from the analytes. In addition, the high resolving power could be combined with cool plasma conditions to improve detection limits for elements like Fe, K, and Ca to <1 ppt. For these reasons, semiconductor users who were looking for the ultimate in performance and were not restricted by financial considerations felt that magnetic sector technology better suited their needs (61).

The other approach, which seems to be gaining a great deal of momentum in the semiconductor community, is to use collision/reaction cell technology. This has been described in greater detail in Chapter 14, "Review of Interferences," but basically uses a combination of collision and reactions to stop the formation of the interfering species before it gets to the analyzer quadrupole. The benefit of this technique over magnetic sector technology for the determination of Fe, Ca, and K is that because high resolving power is not used to reduce the spectral background from the argon-based polyatomic interferences, there is no sacrifice in sensitivity. The background reduction process is achieved by converting the interfering ion into a harmless species by bleeding a reactive gas into the cell to stimulate ion–molecule collisions and reactions. Depending on the design of the cell, the by-products of the reactions and collisions are either removed by mass filtering (62) or kinetic energy discrimination (63). The capabilities of a dynamic reaction cell were recently demonstrated by Collard et al. (64), who showed that method detection limits and spike recovery data for all 21 elements in Grade 5 (10 ppt) hydrogen peroxide could be achieved using strict SEMI methodology. Table 18.9, which was taken from that study, shows typical detection limits and background equivalent concentration (BEC) values achievable with a dynamic reaction cell for all 21 elements defined in a SEMI Grade 5. BEC

TABLE 18.9 Typical Detection Limits and BEC Values Achievable with a Dynamic Reaction Cell for all the 21 Semiconductor-Significant Elements Defined in SEMI Grade 5 (10 ppt) Standard

Element	Detection limit (ppt)	BEC (ppt)
*Aluminum	0.23	0.42
Antimony	0.08	0.08
Arsenic	0.48	1.60
Barium	0.06	0.04
Boron	3.60	7.10
Cadmium	0.08	0.11
*Calcium	0.27	0.63
*Chromium	0.14	0.29
*Copper	0.06	0.68
*Iron	0.49	2.60
Lead	0.07	0.09
Lithium	0.26	0.22
Magnesium	0.23	0.18
Manganese	0.17	0.54
Nickel	0.43	0.66
*Potassium	0.27	2.60
Sodium	0.20	0.22
Tin	0.12	0.88
*Titanium	0.92	1.70
*Vanadium	0.12	0.04
*Zinc	0.63	1.20

Elements with an asterisk were obtained using NH_3 as the reaction gas, while the other elements were determined in standard mode with no reaction gas.
Source: Ref. 64.

values are included because many analysts in the semiconductor community believe that it gives a better indication as to how efficient the background reduction technique is. BEC is defined as the intensity of the spectral background at a particular mass, expressed as a concentration value. The lower the BEC, the easier it is to distinguish an analyte signal from its background.

There appears to be a trend towards the use of collision/reaction technology for the analysis of high-purity semiconductor chemicals because, based on current evidence, the dynamic reaction cell in particular is showing much better performance than has been reported in the literature by other approaches. One of the areas where it is showing enormous potential is in the reduction of polyatomic interferences that cool plasma and high-resolution technology have had significant problems with. For example, the determination of arsenic and chromium in high-purity hydrochloric acid matrix is a very difficult analysis because of the $^{40}Ar^{35}Cl^{+}$ and $^{35}Cl^{16}OH^{+}$ polyatomic spectral interferences on $^{75}As^{+}$ and $^{52}Cr^{+}$, respectively. High-resolution systems do not offer good detection capability because it requires high resolving power to separate $^{75}As^{+}$ from $^{40}Ar^{35}Cl^{+}$ and $^{52}Cr^{+}$ from $^{35}Cl^{16}OH^{+}$, which results in a significant loss of sensitivity. Cool plasma technology has shown limited success for chromium in a high chloride matrix because of matrix suppression effects and is not really suited for arsenic because of its high ionization potential, thus making it very difficult to ionize in a low temperature plasma. On the other hand, Bollinger and Schleisman (65) have demonstrated a detection limit of <2 ppt for arsenic and 7 ppt for chromium in a 10% hydrochloric acid using a dynamic reaction cell. Some of the other successful interference reduction studies reported in the literature using the dynamic reaction cell include $^{40}Ar^{12}C^{+}$ on the determination of $^{52}Cr^{+}$ in an organic matrix (66) and $^{31}P^{16}O^{16}O^{+}$ on the determination of $^{63}Cu^{+}$ in phosphoric acid (67).

I think it is important to emphasize that the semiconductor industry is unique in its demands on instrument manufacturers because unlike any other application area, it is constantly chasing zero. Although this is an unrealistic demand, zero means as little contamination as possible during the manufacturing process, which translates into less defects and therefore a higher yield of semiconductor devices. This is what drives the industry and is reflected in the choice of analytical techniques used. For this reason, any trace element equipment that is applied to contamination control—whether it is the analysis of ultrahigh purity water, the determination of trace metals in chemicals and gases, or carrying out vapor phase decomposition (VPD) studies on the surface of silicon wafers (68)—must be designed specifically for these demands. For example, the surface of the instrument should be smooth as possible, so it does not attract particles of dust. Instrumental components like spray chambers, nebulizers, and pump tubing must be clean and free of contamination. Roughing pumps should be capable of remote operation from the instrument to minimize the effects of vibration. In addition to instrument cleanliness issues, the instrument and sample preparation areas should be housed in at least a Class 1000 clean room, and, for some applications, a Class 100, Class 10, or even a Class 1 might be required. All the volumetric flasks, beakers, storage bottles, etc., need to be of the highest quality with regard to trace metal content. Finally, the calibration standards and acids used to prepare the samples must be the highest purity available. The bottom line is that no matter what type of ICP-MS is used for trace element determinations, even the most sophisticated and high-performance instrument will generate bad data, unless all the sample preparation stages and cleanliness issues are taken into consideration.

NUCLEAR

The types of samples generated by the nuclear industry, including bulk nuclear materials, high/low-level radioactive waste, water soil and biota-based remediation samples, environmental impact studies, and human health monitoring, put unique demands on any analytical technique used for isotopic quantitation. Although traditional ionizing radiation counting techniques have worked exceptionally well over the years, they are painstakingly slow. The inherent problem lies in the fact that to ensure that the radiometric-derived interferences from other sample components are kept to a minimum, time-consuming chemical separations have to be carried out. In addition, the half-life of the analyte isotope has a significant impact on the method detection limit, which means that to get meaningful data in a realistic amount of time, they are better suited for the determination of short-lived radioisotopes. They have been successfully applied to the quantitation of long-lived radionuclides but, unfortunately, require a combination of extremely long counting times and large amounts of sample in order to achieve low levels of quantitation (69).

Limitations in the traditional α (alpha) spectrometry, γ (gamma) spectroscopy, scintillation, and proportional counting technology, especially at extremely low levels, led to the use of atom-counting techniques for radiochemical analysis, such as thermal ionization mass spectrometry (TIMS), secondary ion mass spectrometry (SIMS), accelerator mass spectrometry (AMS), and fission track analysis (FTA). In addition, other techniques were being developed like Fourier transform ion cyclotron resonance (FT-ICR), resonance ionization mass spectrometry (RIMS), and time of flight SIMS (TOF-SIMS), which were primarily being driven by the specific application demands of nuclear-based industry. However, although all these approaches worked very well, depending on the application, they were primarily being used for specific tasks and were not considered truly routine analytical tools. In addition, many of these techniques utilized very complex components, like dedicated nuclear reactors and linear accelerators, which meant they were extremely expensive to manufacture.

The drive in the nuclear industry for a more routine approach that was faster, had less interferences, required easier sample preparation, generated less waste, had good calibration standards, and, very importantly, offered lower cost per sample analysis led researchers to investigate the use of ICP-MS. It was ironic that, although one of the first ICP-MS systems was built at a United States Department of Energy (DOE) site in 1980 (70), the nuclear community was relatively slow in accepting its routine use for radionuclide analysis. However, it soon became clear that the technique was going to be very complimentary to traditional radiation-counting technology used by the nuclear industry (69). This can be seen in Table 18.10, which compares sensitivity, maturity status, typical use, and the advantages/disadvantages of ICP-MS with some of the more established atom-counting techniques and also some of the ones that are still considered to be in the research stage of their development (71).

There is no question that one of the major reasons for the success of ICP-MS in the nuclear industry is that the DOE is changing the mission of many of its facilities from defense-related nuclear materials production to site remediation and monitoring. This change has resulted in a need to fully characterize hazardous wastes and environmental samples, combined with a necessity to routinely monitor workers' exposure to harmful radiation. For this reason, nuclear facilities in the United States and elsewhere are strongly emphasizing these determinations and are demanding better and faster analytical techniques to ensure the quality of the materials that they supply for the production of nuclear energy and other nuclear-related technologies. These factors, which have primarily been driven by DOE initiatives for cost-effective radiochemical analyses, have significantly increased the number of samples from nuclear waste management and nuclear facility cleanups since the mid-1990s, and, as a result, the use and applications of ICP-MS have seen a dramatic increase in this field. Such is the interest in exploring its full potential, that the Nuclear Fuel Cycle Committee (C-26) of ASTM (http://www.astm.org) has put together a Plasma Spectroscopy Task Group (C26-05) to primarily focus on ICP-MS methodology in the nuclear industry. The work of this group is exemplified by the following ICP-MS applications being carried out by the nuclear analytical community.

Applications Related to the Production of Nuclear Materials

Typical analyses carried out in this category include the determination of various radionuclides and the measurement of isotope ratios in enriched

TABLE 18.10 Comparison of Atom-Counting Techniques for the Radionuclides

Technique	Sensitivity	Maturity of technique for radionuclides	Typical use in nuclear industry	Advantages	Disadvantages
TIMS	10^6 atoms	Routine	Isotope ratios for many elements	Quantitative, high precision	Ultraclean sample prep., slow, expensive, interferences from hydrocarbons
FTA	10^6 atoms	Routine	^{239}Pu	Quantitative	Need a nuclear reactor, interference from ^{235}U, expensive
SIMS	10^9 atoms	Routine	Isotope ratios for depth and surface profiling	High spatial resolution and ion imaging	Interferences from hydrocarbons, semiquantitative
AMS	10^5 atoms	Routine	^{10}Be, ^{14}C, ^{26}Al, ^{129}I	10^{15} abundance sensitivity	Complex technology, expensive
ICP-MS	10^6 atoms	Developing/ routine	Isotope ratios for many elements	Rapid, low cost, simple sample preparation	Isobaric and polyatomic spectral and matrix interferences
LAMMA	10^9 atoms	Developing/ routine	Isotope ratios	High spatial resolution	Semiquantitative
FT-ICR	10^9 atoms	Research	Isotope ratios	High resolution, several ion sources	Isobaric and polyatomic spectral interferences
RIMS	10^9 atoms	Research	Isotope ratios	High selectivity, GFAA and GD (glow discharge) ion sources	Nonquantitative

Source: Ref. 71.

uranium compounds like uranium dioxide (UO_2) powder, hydrolyzed uranium hexafluoride (UF_6), and uranyl nitrate liquor (UNL). Depending on the isotopes of interest and the type of mass analyzer used, the problems associated with the analysis of uranium compounds include spectral interferences from actinides and other trace elements in the sample. For example, the determination of ^{99}Tc$^+$ using quadrupole technology can be problematic due to the presence of an isobaric interference from $^{99}Ru^+$ and a molecular interference from $^{98}MoH^+$ (72). If these spectral interferences are not that severe, they can be corrected by mathematical equations; otherwise, some kind of high-resolution mass analyzer must be used to resolve the interfering species away from the analyte isotope. In addition, the high-uranium matrix has the potential to cause severe space charge-induced matrix suppression, especially if low-mass elements are also being determined. To a certain degree, this kind of interference is unavoidable but can be minimized by careful optimization of ion lens voltages to reject the maximum number of uranium ions (73).

If the requirement is for good isotope ratios, double-focusing magnetic sector technology offers the best solution. For example, if the isotope ratio of $^{235}U/^{238}U$ is being monitored in UF_6, it has shown that multicollector (MC) magnetic sector ICP-MS technology will give the best precision data. In fact, this is the preferred methodology over the more traditional TIMS approach because unlike TIMS, the fluoride matrix does not have to be removed. For this reason, the complete analysis using MC-ICP-MS is completed in about one-fifth of the time of TIMS and achieves very similar isotopic ratio precision data (74).

Applications in the Characterization High-Level Nuclear Waste

Some of the many applications carried out in this category include the use of ICP-MS to support the processing, stabilization, and long-term storage of high-level waste (HLW). Common matrices encountered in this kind of work include sludges, slurries, and, in particular, the glass-waste forms that will be used for the isolation of nuclear wastes in underground geological repositories. Analyses usually require the detection of low levels and isotopic content of uranium, in addition to small amounts of actinides and fission products including ^{237}Np (neptunium), $^{239,\ 240}Pu$ (plutonium), ^{241}Am (americium), and ^{244}Cm (curium). The isotopic data for uranium generally do not need to be of the highest accuracy and precision, but to know primarily if the uranium is depleted, natural, or enriched, and if so, an estimate of its enrichment level. These types of samples are further complicated by the fact that they are typically contained in high salt matrices, so they generally have to be diluted to aspirate into the ICP mass spectrometer (75). Other uses for ICP-MS in the application area involve uranium and plutonium solubility studies in groundwater and related samples and also to help determine the efficiency of the separation process when carrying out traditional radiochemical counting. It is also important to point out that because of the dangers associated with characterizing high-level nuclear waste by ICP-MS, most of the work carried out is done with instrumentation that is either completely enclosed in a radiologically controlled glove box or at least with the torch box, sample introduction system, and interface cones positioned inside a radiologically controlled hood (76).

Applications Involving the Monitoring of the Nuclear Industry's Impact on the Environment

Not only is it crucial that the nuclear industry can safely dispose of its low-level waste and monitor its impact on the environment, but also be responsible for cleaning-up old sites related to nuclear power and the production of nuclear weapons. These types of environmental remediation and monitoring activities can generate an unbelievably large numbers of samples. In addition, the samples can be very challenging because typically, the analysis involves the detection of ultratrace levels of radionuclides in complex matrices like soils, groundwaters, sludges, oils, acids, and organic wastes. For that reason, sample preparation steps involving matrix removal and/or analyte preconcentration are often going to be required to carry out successful determinations at such low levels. However, the large number of samples generated also requires an analytical method that is fast and fully automated.

An example of this approach was reported by Hollenbach et al. (77) who determined a suite of radionuclides (^{230}Th, ^{234}U, ^{239}Pu and ^{240}Pu) with relatively long half-lives in series of soil samples using automated flow injection (FI) to separate the matrix and preconcentrate the analytes by the use of solid phase extraction and pass the eluent directly into the ICP-MS nebulizer. Detection capability in solution (becquerel/liter—Bq/L) and directly in the soil (becquerel/kilogram—Bq/kg) using this methodology is shown in Table 18.11.

The conclusion of the authors was that this fully automated quadrupole ICP-MS method was faster, less labor-intensive, and generated less laboratory waste than traditional radiochemical methods. They also pointed out

TABLE 18.11 Detection Capability of a Group of Radionuclides in Soil Using FI-ICP-MS

Isotope	D/L in solution (Bq/L)	D/L in soil (Bq/kg)
^{230}Th	0.03	3.0
^{234}U	0.006	0.6
^{239}Pu	0.004	0.4
^{240}Pu	0.02	2.0

Source: Ref. 77.

that an additional benefit was that they were able to detect radionuclides like ^{239}Pu and ^{240}Pu, which could not be resolved by traditional radiochemical methods.

Applications Involving Human Health Studies

Protecting its workers from the long-term effects of exposure to uranium compounds is an extremely important role of the nuclear industry. This means that both the quality of the air and the workers themselves must be monitored for uranium and if required, other radionuclides, on a regular basis. Air monitoring usually means sampling the air using personal monitors or at various locations in and around the nuclear facility, while monitoring the workers normally means taking blood and urine samples over a period of time and looking for a trend or pattern in the data of the analytes of interest. As you can imagine, uranium levels, particularly of the less abundant isotopes like ^{234}U and ^{235}U, are going to be extremely low. So not only are the detection levels a problem by traditional techniques, but also in the case of air filters and blood, the amount of sample for analysis can be limited. When this is combined with the fact that routine environmental monitoring generates a large number of samples, it makes ICP-MS ideally suited for these kinds of human health studies. In fact, a study done in the mid-1990s showed that ICP-MS is capable of generating good-enough uranium isotope ratio data on air filters in less than 12 min, with as little as 10 ng of uranium, in order to identify the source of a particulate effluent from a nuclear operation (78).

In another study carried out at a different nuclear establishment, the isotopes ^{238}U and ^{235}U were determined (along with calculated ^{234}U concentrations) in a series of human urine samples using a concentrated aqua regia wet oxidation method to dissolve the uranium and destroy the organic matter. The uranium was selectively separated from the matrix using anion exchange, eluted with dilute nitric acid, and then aspirated into the ICP mass spectrometer. Using this method, a detection limit of 6 ng/L was achieved, with excellent spike recoveries at the 200-ng/L level, which met both plant and industry standard (ANSI 13.30) internal dose assessments for total uranium (79).

OTHER APPLICATIONS

All the application segments discussed up to now are predominantly utilizing ICP-MS in a routine manner using well-established methods. As a result, environmental, biomedical, geochemical, semiconductor, and nuclear represent over 80% of all applications being carried out by ICP-MS today. However, there are a number of other industries that rely on ICP-MS, not so much as a high-throughput tool, but focus more on its use as a research-oriented technique to solve problems associated with a manufacturing process or perhaps to analyze difficult samples that other trace element techniques cannot handle. In addition, there are many academic institutions that use ICP-MS for fundamental research into ionization mechanisms, plasma diagnostics, sample introduction methods, and mass separation resolution studies. Let us take a closer look at some of these applications.

Metallurgical

The metallurgical industry uses a number of mature and well-established techniques like x-ray fluorescence (XRF), arc/spark optical emission (OES), and glow discharge optical emission spectrometry (GDOES) to support quality control aspects of the refining and production of metals such as iron, steel, aluminum, copper, nickel, and various alloys that are used in the manufacture of cars, ships, aircraft, and related industries. In addition, other more "exotic" techniques like secondary ion mass spectrometry (SIMS), glow discharge mass spectrometry (GDMS), and laser microprobe mass spectrometry (LAMMS) are used if lower detection levels are needed or if there is a requirement for the analysis of defects or inclusions on the surface of these materials. However, the one thing that all these techniques have in common, which makes them very attractive to a manufacturer of high-temperature alloys used by the aerospace industry or low-carbon steel strip made specifically for the auto industry, is that they are solid-sampling techniques. In other words, a multielement analysis can be carried out with very little or no sample preparation.

For this reason, there has generally been less demand for the multielement analysis of solutions in the metallurgical industry. Usually, it was only required if there were some elemental heterogeneity or segregation problems with the sample itself or if there was a need to confirm an abnormal result generated by one of the solid-sampling techniques. In these situations, the sample had to be dissolved in some acid medium and be analyzed by either flame AA, if only a few elements were required, or ICP-OES, if many elements were needed. Only in extreme cases when the analytes in solution were below AA/ICP-OES detection limits would GFAA be required. For all these reasons, there was no real demand for ICP-MS in the metallurgical industry, not because it was not a suitable technique, but because most of the trace element determinations in the industry were being adequately addressed by the other, well-established approaches.

However, over the past 5–10 years, we are beginning to see a growing trend in the use of ICP-MS in this application segment (80). This is partly driven by the fact that high-purity metals and complex alloys are often very challenging to analyze by emission or absorption-based techniques, like AA, ICP-OES, or GD-OES, because of spectral and matrix interferences generated by the high levels of major elements in the sample (81). This has a major impact on detection capability, especially for the aircraft and aerospace industries, which use very high-purity metals and high-temperature alloys. However, I believe the major reason for the recent growth in ICP-MS in the metallurgical industry is because of the exciting potential of coupling laser ablation with ICP-MS. It is clear that modern 266-nm Nd:YAG laser ablation systems, especially the ones optimized for bulk analysis, are now capable of ablating just about any metal and producing a continuous stream of fine particles suitable for an ICP-MS system. As a result, LA-ICP-MS is not only offering metallurgical chemists the ability to directly analyze solid samples with good stability and precision, but also having the flexibility to determine ultratrace levels in solutions with far superior detection capability than FAA, GFAA, or ICP-OES. In fact, for many types of samples, its performance is as good as GD-MS. This is demonstrated in Table 18.12, which shows the determination of a group of elements in a Ni/Mo/W high-temperature alloy by both techniques (51). It should be emphasized that the aim is just to show a comparison of the data and is not intended to be an evaluation of accuracy by comparing it with a certified reference material. Results for all elements are expressed as parts per billion in the solid.

The added benefit of LA-ICP-MS is that the sampling area can be as low as 10μ, so by rastering across the surface, it can also detect any heterogeneity or segregation on the surface of the sample. This sort of sampling precision is beyond the capability of GD-MS because it is used predominantly as a bulk sampling technique. In fact, elemental segregation on the surface of the

TABLE 18.12 The Determination of a Group of Elements in an A1/Mo/W High-Temperature Alloy by GD-MS and LA-ICP-MS

Element	GD-MS (ppb)	LA-ICP-MS (ppb)
Na	0.14	0.08
Mg	78	79
Si	323	255
Zr	354	314
Nb	218	170
Sn	1.4	1.1
Hf	125	110

Source: Cetac Technologies.

sample could be the reason why the data in this table do not agree for all elements in the alloy.

Petrochemical and Organic-Based Samples

In the production of petrochemicals and related products, it is critical for refineries and chemical plants to closely monitor trace element contamination levels at various stages of the manufacturing process. For example, in the refining of crude oil, some elements such as Ni and V, even at "parts per billion" levels, can act as catalyst poisons and cause enormous problems owing to the volumes of hydrocarbons that are processed (82). In addition, if the final product is intended for use by the food industry or the manufacture of electronic devices, the specifications for trace element contamination are even more stringent.

The problem is that the analysis of petrochemical samples can be extremely difficult because of the complex nature of crude oils, distillates, residues, fuel oils, petroleum products, organic solvents, and all the various by-products from refining crude oil. These complex oil-based samples pose major problems for any analytical technique because of the difficulty in introducing them directly into the instrument. So the analytical challenge for any trace element technique being used in the petrochemical industry is to be able to carry out fast, reliable determinations of total and also speciated forms of critical metals, in a wide variety of complex samples, with the minimum of sample preparation.

Unfortunately, some of the traditional ways getting petrochemical samples into solution are extremely slow and labor-intensive. Common sample preparation methods include digestion with strong acids/oxidizing agents and/or ashing the sample in a muffle furnace and redissolving the residue in a suitable solvent. The acid digestion procedure alone tends to lead to an incomplete dissolution because of the high level of carbonaceous material, so for that reason, the ashing procedure or a combination of oxidation and ashing is preferred. This also allows for preconcentration of the sample to provide adequate amounts of the test analytes to be analyzed, if they are present at ultratrace levels. The choice of which of these traditional sample preparation approaches to use is often determined by the final instrumental technique. However, they all have a number of things in common—apart from taking up a considerable amount of time in the total analytical process, they can also lead to loss of sample, loss of volatile analytes, and major contamination problems.

Avoiding problems like these were among the reasons why the petrochemical community became really interested in ICP-MS a number of years ago. Previously, ICP-OES was one of the preferred techniques for the multielement analysis of oil-based samples. However, because of the achievable detection limits of ICP-OES, a sample preparation technique known as the sulfated ash method (SASH) usually had to be used (83). This approach, which involved oxidation of the oil sample with concentrated sulfuric acid and high-temperature ashing, took approximately 3 days to complete to make sure all the analytes were in solution. When the use of ICP-MS was investigated, they found that because of its extremely high sensitivity, a simple dilution of the sample with a solvent like toluene could be used. In other words, the lengthy sulfated ash method used to get the analytes into solution could be avoided, which represented an enormous time saving. Unfortunately, there was a slight downside to the ICP-MS methodology. In order to directly aspirate the toluene-diluted oil sample, a special chilled spray chamber has to be used to desolvate the sample. In addition, high RF power is necessary together with a small amount of oxygen in the nebulizer gas flow to "burn off" any remaining solvent. This has the effect of stopping carbon deposits building up on the interface cones and also minimizing the formation of carbon-based spectral interferences. However, once the instrument is set up and optimized with this sample introduction system, the analysis of most organic-based samples is relatively straightforward. Besides the enormous time saving, contamination problems are dramatically reduced and the loss of volatile elements is avoided, compared to the complex SASH sample preparation procedure. Table 18.13 compares ICP-OES using the sulfated ash method and ICP-MS using a simple 1:1000 dilution in toluene for the determination of Ni and V in NIST 1618 certified reference fuel oil (84). It should be noted that large dilutions are typical for the analysis of oil-based samples by ICP-MS or ICP-OES in order to minimize sample transport and viscosity effects. It can be seen that the accuracy and the precision of both

TABLE 18.13 The Determination of Ni and V in NIST 1618 Certified Reference Fuel Oil by ICP-OES Using the Sulfated Ash Method and ICP-MS Using a Simple 1:1000 Dilution in Toluene

NIST 1618 CRM	Total sample preparation/ analysis time	Sample weight (g)	Ni (ppm)	RSD (%)	V (ppm)	RSD (%)
ICP-OES/SASH	72 hr	5	76.2	1.7	426	1.3
ICP-MS/dilution	45–60 min	3	75.9	1.5	424	1.9
Certificate value	—	—	75.2 ± 0.4	—	423.1 ± 3.4	—

Source: Ref. 84.

methods are similar and in good agreement with the certificate, but the ICP-MS determination is almost 100× faster.

Food Analysis

The trace element analysis of foodstuff has always been important because the nature and the concentration of many elements are related to the biological role they play in the physiology (biological study of the functions) of the living organism. Factors that influence trace element levels in food materials include natural processes, inadvertent contamination during growth and manufacturing, and preparation processes. Some elements like As, Cd, Hg, and Pb are considered toxic, while others like Se, Cr, Zn, Mn, and Ni have a dual personality because in some forms, they are essential, and in other forms, they are toxic. Therefore there is a need to classify two groups of trace elements in foodstuffs—toxic elements, which are typically present at trace levels, and nutritional elements, which are mostly, but not exclusively, present at higher levels. Therefore the challenge of any technique used in the food industry is not only to be able to determine ultratrace levels (sub-parts per billion), but also be able to determine higher concentration levels (typically parts per million). This has traditionally been done by a combination of FAA, GFAA, and ICP-OES, but clearly, if many elements need to be classified, it can be very time-consuming, especially if conventional acid digestion methods are used to get the sample into solution.

For these reasons, ICP-MS has proved to be a very attractive option for the analysis of foodstuff, especially as most modern instruments now have the capability to extend the dynamic range to determine much higher levels. There have been a number of publications on the use of ICP-MS for the analysis of foodstuffs (85,86), but they mainly focused on elements at the trace level because earlier technology was not able to handle such a wide spread in analyte concentrations with one sample preparation method. However, we are now beginning to see more applications in the open literature on the multielement analysis of food, at both high and low levels, using a single sample preparation. For example, Zhou and Liu (87) showed that 15 elements (V, Cr, Mn, Co, Ni, Zn, As, Se, Mo, Pd, Cd, Sn, Hg, Tl, Pb, Rh, Re), from low parts per trillion to high parts per billion levels, could be determined in 16 varieties of foodstuff with good accuracy and precision using a simple external calibration. The benefit of this methodology is that all elements can be measured at the same time in one solution, prepared by digesting the sample with concentrated nitric acid in a microwave oven. This is exemplified in Table 18.14, which shows the determination of a group of selected elements in various food-based Chinese CRMs (National Research Center for Certified Reference Materials, Beijing, China) (87). All results are expressed in ng/g in the food.

TABLE 18.14 Determination of a Group of Selected Elements (in ng/g in Food) in Various Food-Based Chinese Certified Reference Materials

	Rice (ng/g)		Pork liver (ng/g)		Mussels (ng/g)	
Element	Found	Cert.	Found	Cert.	Found	Cert.
Mn	9.5	9.8	9.37	8.32	10.7	10.2
Co	—	—	—	—	1.18	0.94
Ni	—	—	—	—	0.90	1.03
Zn	14.8	14.1	180	172	136	138
As	0.051	0.051	0.066	0.044	5.5	6.1
Se	0.050	0.045	—	—	—	—
Mo	—	—	—	—	0.62	0.6
Cd	0.018	0.020	0.077	0.067	4.0	4.5
Hg	—	—	—	—	0.073	0.067
Pb	—	—	0.59	0.54	—	—

Source: Ref. 87.

In addition to carrying out the total metal content of food-related samples, ICP-MS coupled with various chromatography separation devices is proving an invaluable detection technique to characterize extremely low levels of various elemental species in foodstuffs. An example of using ICP-MS in this way was exemplified in a recent study into the ability of selenium as an anticarcinogen (88). Se is both an essential and a toxic element. On the one hand, it is thought to have anticancer properties, by preventing cell membranes from damage due to oxidation. On the other hand, a selenium deficiency causes skeletal and cardiac muscle dysfunction, while at high levels, some forms of selenium are considered extremely toxic. It is therefore very important to know the biodegradation process of different selenium compounds in plants, like garlic, onions, and broccoli, in order to get a better understanding of their anticancer properties. The research groups in this study used ICP-MS in conjunction with HPLC to separate various organo-selenium compounds in plant material. They showed that trace levels of selenoamino acids, including selenocysteine, selenomethionine, methylselenocysteine, and propylselenocysteine, could be determined even in the presence of large amounts of sulfur. This is particularly significant because selenium predominantly follows the chemistry of sulfur, which can present considerable separation challenges by traditional analytical techniques. Fortunately, using ICP-MS, sulfur does not pose any serious problems in the determination of selenium.

However, it should be emphasized that the determination of selenium is not that easy by quadrupole ICP-MS because of a significant spectral interference from the argon dimer $^{40}Ar^{40}Ar^+$ on the major selenium isotope at mass 80. For that reason, a less abundant isotope has to be used for quantitation, which unfortunately degrades detection capability (D/L 50–100 ppt). High-resolution offers approximately 5–10× better performance but still has to use one of the less abundant isotopes because it requires extremely high resolving power to separate ^{80}Se from both $^{40}Ar^{40}Ar^+$ and $^{80}Kr^+$ (krypton is an impurity in the argon gas), which has a dramatic effect on sensitivity. These limitations have led researchers to investigate the use of collision/reaction cell technology to eliminate the formation of the argon dimer and to determine selenium at mass 80—its major isotope. In fact, on the evidence presented to date, this technology might offer the best approach to determine selenium at sub-parts per trillion levels (89). This is particularly relevant in speciation studies because individual organoselenium compounds will be significantly less than the total selenium concentration in the plant-, vegetable-, or food-related material.

SUMMARY

The applications discussed in this chapter account for over 95% of all applications being carried out by ICP-MS. However, there are a number of other groups and industries that also use the technique but were not included in the pie chart because individually, they might only represent a few percent of the total market. Some of these applications include Forensic Science (90), Polymers and Plastics (91), Pharmaceuticals (92), Ceramics (93), and Pottery (94). These market segments will not be discussed in this chapter, but if you are working in one of these areas and want to know more about the capabilities of ICP-MS, I suggest you read one of the cited references.

The aim of this chapter was to give a "flavor" of the application potential of ICP-MS and a better understanding of why it is the fastest-growing trace element technique available today. If there is one common theme that runs through many of these applications, it is the unparalleled detection limits it has to offer. However, when this is combined with its rapid multielement characteristics, isotopic measurement capability, freedom from interferences, and ease of use, it is clear that it is only a matter of time before ICP-MS becomes the dominant technique for trace metal determinations.

FURTHER READING

1. EPA Method 200.8: December 5, 1994. Federal Register-Vol. 59 [232] p. 62546.
2. Federal Register-Vol. 66, No. 14, January 22, 2001/Rules and Regulations.
3. EPA Method 200.8: January 27, 2000. Federal Register-Vol. 65 [18] p. 4360.

4. Method 1638—Determination of trace elements in ambient waters by ICP-MS—EPA 821-R-95-031, April 1995.
5. Method 1640—Determination of trace elements in ambient waters by on-line chelation—EPA 821-R-95-033, April 1995.
6. Method 1669—Sampling ambient waters for the determination of trace metals in EPA quality criteria levels and quality control supplement—EPA 821-R-95-034, April 1995.
7. SW-846 Method 6020 (RCRA Programs): January 13, 1995. Federal Register-Vol. 60 [009] p. 3089—Update II of third edition.
8. EPA Method 6020A: January 1998.
9. EPA Inorganic Statement of Work, December 2001. Document ILM05-2.
10. Lobinski R, Pereiro IR, Chassaigne H, Wasik A, Szpunar J. J Anal At Spectr 1998; 13:860–867.
11. Wolf RE, Grosser ZA. At Spectr 1997; 18(5):145–151.
12. McLaren JW, Beauchemin D, Berman SS. Anal Chem 1987; 59:610.
13. Meyers Robert A, ed. Encyclopedia of Environmental Analysis and Remediation 37. New York: John Wiley and Sons, 1998:23–2356.
14. Wolf RE, Grosser ZA. Am Environ Lab, February 1997; 36–40.
15. Beauchemin D, McLaren JW, Mykytiuk AP, Berman SS. Anal Chem 1987; 59:778.
16. May TW, Wiedmeyer RH, Brumbaugh WG, Schmitt CJ. At Spectr 1997; 18(5):133–139.
17. Ting BTG, Janghorbani M. Anal Chem 1986; 58:1334.
18. Lyon TDB, Fell GS, Hutton RC, Eaton AN. J Anal At Spectr 1988; 3:601.
19. Barnes RA. Anal Chim Acta 1993; 283:115.
20. Nixon DE, Moyer TP. Spectrochim Acta, Part B 1996; 51:13–25.
21. Vanhoe H, Vandecasteele C, Verieck J, Dams R. Anal Chem 1989; 61:1851.
22. Brunk S. At Spectr 1994; 15(4):145–149.
23. Outridge PM, Hughes RJ, Evans RD. At Spectr 1996; 17(1):1–8.
24. Bortoli A, Gerotto M, Marchiori M, Palonta R, Troncon A. J Microchem Anal 1992; 46:167.
25. Pruszkowski E, Neubauer K, Thomas R. At Spectr 1998; 19(4):111–115.
26. Shibata Y, Morita M. Anal Sci 1989; 5:107.
27. Beauchemin D, Siu KWM, McLaren JW, Berman S. J Anal At Spectr 1989; 4:285.
28. Crews HM, Dean JR, Ebdon L, Massey RC. Analyst 1989; 114:895.
29. Gercken B, Barnes RM. Anal Chem 1991; 63:283.
30. Vela NP, Caruso JA. J Anal At Spectr 1996; 11:1129.
31. Majidi V, Miller-Ihli NJ. Analyst 1998; 123:803.
32. Date AR, Gray AL. Analyst 1983; 108:159.
33. Haskin LA, Wilderman TR, Haskin MA. Radioanal Chem 1968; 1:337–348.
34. Walder AJ, Freeman PA. J Anal At Spectr 1992; 7:571–575.
35. Jarvis KE, Jarvis I. Geostand Newsl 1988; 12(1).
36. Figueiredo AMG, Marques LS. Geochim Bras 1989; 3(1).
37. Pruszkowski E, Barrett P. Spectrochim Acta 1994; 39B:485.

38. Lichte FE, Meier AL, Crock JG. Anal Chem 1987;59(8):1150–1157.
39. Douglas DJ. Can J Spectrosc 1989; 34(2):38–49.
40. Stroh A, Voellkopf U, Denoyer ER. J Anal At Spectr 1992; 7:1201–1205.
41. Denoyer ER, Ediger R, Hager J. At Spectr 1989;10(4):97–102.
42. Beamish FE, Van Loon JC. Analysis of Noble Metals. New York: Academic Press, 1977:178.
43. Riddle C, Vander Voet A, Doherty W. Geostand Newsl 1988; 12(1):203.
44. Longerich HP, Jenner GA, Jackson SE. Chem Geol 1990; 83(105).
45. Dickin AP. Radiogenic Isotope Geology. Cambridge, UK: Cambridge University Press, 1995.
46. Halicz L, Erel Y, Veron A. At Spectr 1996; 17(5):186–189.
47. Rosman KJR, Chisholm W, Boutron CF, Candelone JP, Gorlach U. Nature 1993; 362:333.
48. Vanhaecke F, Moens L, dams R, Taylor P. Anal Chem 1996; 68(3):567–569.
49. Halliday AN, Lee DD, Christensen JN, Rehkamper M, Yi W, Luo X, Hall CM, Ballentine CJ, Pettke T, Stirling C. Geochim Cosmochim Acta 1998; 62:919–940.
50. Vanhaecke F, Moens L, Tanner SD, Baranov VI, Bandura DR. PerkinElmer Sciex Application Note, Chemical Resolution of 87Rb/87Sr Isobaric Overlap: Fast Rb/Sr Geochronology by Means of DRC ICP-MS, D-6538, 2001.
51. Howe T, Shkolnik J, Thomas R. Spectroscopy 2001; 16(2):54–66.
52. Jackson SE, Longerich HP, Dunning GR, Fryer BJ. Can Mineral 1992; 30:1049–1064.
53. Gunther D, Heinrich CA. J Anal At Spectr 1999; 14:1369.
54. Gonzalez J, Mao XL, Roy J, Mao SS, Russo RE. J Anal At Spectr 2002; 17:1108–1113.
55. Shuttleworth S, Kremser D. J Anal At Spectr 1999; 13:697–699.
56. Mahoney PP, Li G, Hieftje GM. J Anal At Spectr 1996; 11:401–405.
57. Gunther D, Horn I, Hattendorf B. Fresenius' J Anal Chem 2000; 368:4–14.
58. International Technology Roadmap for Semiconductors (ITRS) http://www.public.itrs.net, 2001.
59. Book of SEMI Standards (BOSS). San Jose, CA: Semiconductor Equipment and Materials International.
60. Sakata K, Kawabata K. Spectrochim Acta 1994; 49B:1027.
61. Chang SJ, Chen SL. Instrum Today 1998; 20:51–58.
62. Tanner SD, Baranov VI. At Spectr 1999; 20(2):45–52.
63. Feldmann, Jakubowski N, Thomas C, Stuewer D. Fresenius' J Anal Chem 1999; 365:422–428.
64. Collard JM, Kawabata K, Kishi Y, Thomas R. Micro, January 2002; 39–46.
65. Bollinger DS, Schleisman AJ. At Spectr 1999; 20(2):60–63.
66. Neubauer K, Voellkopf U. At Spectr 1999; 20(2):64–68.
67. Kishi Y, Kawabata K, Thomas R. Spectroscopy 2003; 18:1.
68. Radle M, Lian H, Nicoley B, Howard AJ. Semicond Int, July 2001.
69. Morrow RW, Crain JS, eds. Applications of Inductively Coupled Plasma to Radionuclide Determinations. West Conshohocken, PA: ASTM, 1995.

70. Houk RS, Fassel VA, Flesch GD, Svec HJ, Gray AL, Taylor CE. Anal Chem 1980; 52:2283.
71. Counsel of Ionizing Radiation Measurements Workshop on Standards, Intercomparisons, and Performance Evaluations for Low-level and Environmental Radionuclide Mass Spectrometry—Meeting Proceedings. Gaithersburg, MD: NIST, April 1999.
72. Makinson PR. Morrow RW, Crain JS, eds. Applications of ICP-MS to Radionuclide Determinations. West Conshohocken, PA: ASTM, 1995:7–19.
73. Denoyer ER, Jacques D, Debrah E, Tanner SD. At Spectr 1995; 16(1):1.
74. Walder AJ, Hodgson T. Morrow RW, Crain JS, eds. Applications of ICP-MS to Radionuclide Determinations. West Conshohocken, PA: ASTM, 1995:20–25.
75. Kinard WF, Bibler NE, Coleman CJ, Dewberry RA, Boyce WT, Wyrick SB. Morrow RW, Crain JS, eds. Applications of ICP-MS to Radionuclide Determinations. West Conshohocken, PA: ASTM, 1995:48–58.
76. Barrero Moreno JM, Betti M, Garcia Alonso JI. J Anal At Spectr 1997; 12:355–361.
77. Hollenbach M, Grohs J, Mamic S, Koft M. Morrow RW, Crain JS, eds. Applications of ICP-MS to Radionuclide Determinations. West Conshohocken, PA: ASTM, 1995:99–115.
78. Price-Russ G III, Bazan JM. Morrow RW, Crain JS, eds. Applications of ICP-MS to Radionuclide Determinations. West Conshohocken, PA: ASTM, 1995:131–140.
79. Vita OA, Mayfield KC. Morrow RW, Crain JS, eds. Applications of ICP-MS to Radionuclide Determinations. West Conshohocken, PA: ASTM, 1995:141–147.
80. Kuss HM, Bossmann D, Muller M. Nauche R, ed. Proceedings of the 3rd International Conference on Progress of Analytical Chemistry in the Iron and Steel Industry (EUR14113 EN), 1992:302–307.
81. Kuss HM, Bossmann D, Muller M. At Spectr 1994; 15(6):148–150.
82. Botto RI, Zhu JJ. J Anal At Spectr 1994; 9:905.
83. Sulfated Ash Sample Preparation Method—ASTM Method D-874.
84. McElroy F, Mennito A, Debrah E, Thomas R. Spectroscopy 1998; 13(2):42–53.
85. Crews HM. Int Lab 1993; 23:38.
86. Sheppard BS. Analyst 1994; 119:1683.
87. Zhou H, Liu J. At Spectr 1997; 18(4):115–118.
88. Uden P, Tyson J, Kotrebai M, Block E. Paper No. 870. FACSS Conference, Vancouver, BC.
89. Neubaur K, Wolf RE. PerkinElmer Sciex Application Note, Low Level Selenium Determination, D-6358, 2000.
90. Koons RD. J Forensic Sci 1998; 43(4):748–754.
91. Wolf RE, Thomas C, Bohlke A. Appl Surf Sci 1998; 127–129:299–303.
92. Wolf RE. At Spectr 1997; 18(6):169–174.
93. Bettinelli M, Baroni U, Bilei F, Bizzarri G. At Spectr 1997; 18(3):77–79.
94. Chaudhary-Webb M, Pascal DC, Elliott WC, Hopkins HP, Ghazi AM, Ting WC, Romieu I. At Spectr 1998; 19(5):156–163.

19

Comparing ICP-MS with Other Atomic Spectroscopic Techniques

Now that we have presented the basic principles of ICP-MS and its major application strengths, let us turn our attention to comparing it with other approaches to trace element analysis. ICP-MS is a very powerful technique, but is it the right one for your laboratory? Do you need its multielement capability? Are the detection limits of your current techniques, good enough? Will your operators be able to handle the more complicated method development of ICP-MS? Are you prepared for its increased running costs? In other words, have you considered the implications of owning an ICP mass spectrometer? To help you answer these questions, Chapter 19 will take a look at the strengths and weaknesses of ICP-MS and compare them with those of other trace element techniques such as flame atomic absorption (FAA), electrothermal atomization (ETA), and inductively coupled optical emission spectrometry (ICP-OES), in order to help you decide if ICP-MS is really a good fit for your laboratory. (This chapter has been adapted from two articles I wrote for Today's Chemist at Work magazine (1,2) and has been used with permission from the American Chemical Society.)

Since the introduction of the first commercially available atomic absorption spectrophotometer (AAS) in the early 1960s, there has been an increasing demand for better, faster, higher performance, easier-to-use, and more flexible trace element instrumentation. A conservative estimate shows that today's marketplace for atomic spectroscopy (AS)-based instruments such as atomic absorption, inductively coupled plasma optical emission (ICP-OES), and inductively coupled plasma mass spectrometry (ICP-MS) represents over $500M in annual revenue. As a result of this growth, we have seen a rapid emergence of more sophisticated equipment and easier-to-use software. When this is combined with an increase in the number of manufacturers of both instrumentation and sampling accessories, the choice of which technique to use is often very unclear.

In order to select the best technique, for a particular analytical problem, it is important to understand exactly what the problem is and how it is going to be solved. For example, if the requirement is to monitor copper at percentage levels in a copper plating bath and it is only going to be done once per shift, it would be inappropriate to choose a rapid ultratrace multielement technique such as ICP-MS. A single element technique such as FAA would probably suffice for this application. Although this might be an exaggerated example, it emphasizes that there is an optimum atomic spectroscopic technique for every application problem. When choosing a technique, it is important to understand not only the application problem, but also the strengths and weaknesses of the technology being applied to solve the problem. However, there are many overlapping areas between the major atomic spectroscopy techniques, so it is highly likely that for some applications, more than one technique would be suitable. For that reason it is important to go through a carefully thought-out evaluation process when selecting a piece of equipment.

First of all, let us take a brief look at the most commonly used atomic spectroscopy techniques—atomic absorption, ICP optical emission, and ICP mass spectrometry. There are different variations of each technique, but basically atomic absorption uses the principle of generating free atoms (of the element of interest) in a flame or electrothermal atomizer (sometimes referred to as graphite furnace or GFAA) and measuring the amount of light absorbed from a wavelength specific light source. Inductively coupled plasma emission uses the principle of exciting atoms in a plasma and measuring the amount of light the atoms emit when they fall back down to a ground (stable) state. And, as we have discussed in the previous chapters, ICP mass spectrometry uses the plasma to generate ions and measures the number of ions produced at a particular mass-to-charge ratio. A simple schematic of atomic absorption, emission, fluorescence, and mass spectrometry is shown in Figure 19.1.

Although atomic fluorescence is considered an atomic spectroscopic technique, it will not be covered in this chapter. Let us take a look at the other AS techniques in greater detail.

FLAME ATOMIC ABSORPTION

This is predominantly a single-element technique that uses a flame to generate ground-state atoms. The sample is aspirated into the flame via a nebulizer and a spray chamber. The ground-state atoms of the sample absorb light of a particular wavelength from an element-specific, hollow cathode lamp source.

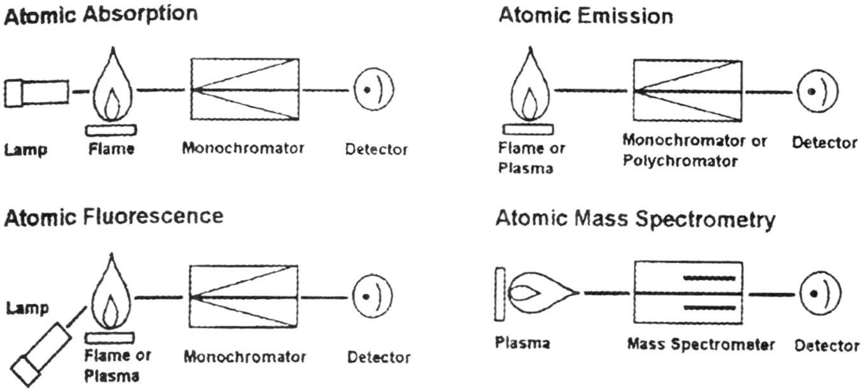

Atomic Absorption

Lamp   Flame   Monochromator   Detector

Atomic Emission

Flame or   Monochromator or   Detector
Plasma   Polychromator

Atomic Fluorescence

Lamp

Flame or   Monochromator   Detector
Plasma

Atomic Mass Spectrometry

Plasma   Mass Spectrometer   Detector

FIGURE 19.1 Simple schematic diagram of the principles of atomic absorption, emission, fluorescence, and mass spectrometry. (Courtesy of PerkinElmer Life and Analytical Sciences.)

The amount of light absorbed is measured by a monochromator (optical system) and detected by a photomultiplier or solid-state detector, which converts the photons into an electrical pulse. This absorbance signal is used to determine the concentration of that element in the sample. Flame AA typically uses about 2–5 mL/min of liquid sample and is capable of parts-per-million (ppm) detection limits.

ELECTROTHERMAL ATOMIZATION

This is also mainly a single-element technique, although multielement instrumentation is now available. It works on the same principle as FAA, except that the flame is replaced by a small heated tungsten filament or graphite tube. The other major difference is that a very small sample (typically 50 μL) is injected automatically onto the filament or into the tube and not aspirated via a nebulizer and a spray chamber. Because the ground-state atoms are concentrated in a smaller area than a flame, more absorption takes place. The result is that ETA offers about 100× lower detection limits than FAA.

RADIAL-VIEW ICP OPTICAL EMISSION

ICP-OES is a multielement technique that uses a traditional radial (side-view) inductively coupled plasma to excite ground-state atoms to the point where they emit wavelength-specific photons of light, characteristic of a particular element. The number of photons produced at an element-specific wavelength is measured by high resolving-power optics and photon-sensitive detection system. This emission signal is directly related to the concentration of that element in the sample. The analytical temperature of an ICP is about 6000–7000 K, compared to a flame, which is typically 2500–4000 K. A radial ICP can achieve similar detection limits to FAA, for the majority of elements, but has the advantage of offering much better performance for the refractory and rare earth elements. Sample requirements for ICP-OES are approximately 1 mL/min.

AXIAL-VIEW ICP OPTICAL EMISSION

The principle is exactly the same as radial ICP-OES, except the plasma is viewed horizontally (end-on). The benefit is that more photons are seen by the detector and, as a result, detection limits can be as much 2–10× lower, depending on the design of the instrument. The disadvantage is that more severe matrix interferences are observed with an axial ICP. Sample requirements are the same as for radial ICP-OES.

INDUCTIVELY COUPLED PLASMA MASS SPECTROMETRY

This has been described in great detail in the previous chapters. The fundamental difference between ICP-OES and ICP-MS is that in ICP-MS, the plasma is not used to generate photons but to generate positively charged ions. The ions produced are transported and separated by their atomic mass-to-charge ratio, using a mass-filtering device such as a quadrupole. The generation of such large numbers of positively charged ions allows ICP-MS to achieve detection limits at the part-per-trillion (ppt) level compared to ICP-OES, which is typically in the ppb range.

This is not meant to be a detailed description of the fundamental principles of each technique, but a basic understanding as to how they differ from each other. To begin the process of deciding whether ICP-MS is the best technique for your needs, there are basically four steps to consider (1):

- Define the analytical objective.
- Establish selection criteria.
- Define the application tasks.
- Compare the techniques.

Each step in the process should serve to focus attention on the technique(s) that best meet the requirements of the analytical task. Let us take a closer look at these steps.

Define the Objective

In this step, the analytical objective should be broadly defined. For example, what is the concentration of iron in high-purity hydrochloric acid or how much arsenic is in contaminated soil? However, it is important not to lose sight of what one is actually trying to accomplish with this analysis. In other words, what decisions will be made based on knowing the trace element composition of the sample. Before proceeding to specifics, one should have a simple view of what, at the end of a complex evaluation of several different analytical techniques, is the desired result. Once that has been done, one can proceed and focus in on the techniques that could possibly accomplish this task.

Establish Criteria

Use this process to focus in on the right techniques. The field should now be narrowed down to establish a set of practical criteria, which might eliminate some of the less suitable techniques for a particular application. Some of these criteria will include, but are not limited to, instrument reliability, quality of data, sample throughput capability, ease-of-use, operator training requirements, or availability of application material.

Define the Application Task

By rigorously defining the task, it will become relatively clear what techniques to evaluate. By comparing and contrasting the attributes of each of the techniques, one can begin to appreciate the value of each one and determine how the instrumentation will be used in the laboratory. The factors/issues that influence this decision will vary depending on the individual situation. They may not all be valid, but some will be of more importance than others. However, before an informed decision can be reached, each one should be considered to some degree. These issues can be broken down into four major categories—application, installation, user, or financial considerations. Let us take a closer look at them.

Application

This will include information about the elemental requirements; what detection limits and concentration ranges are expected and what accuracy and precision are required. It will also include sample information, such as how many samples are expected and at what frequency; how much time can be spent on sample preparation and how quickly must they be analyzed. Sometimes the amount of available sample will dictate the selection or whether interferences from the matrix components have a major impact on the analysis.

Installation

Installation factors might include the size of the instrument and how much lab space is required, what services are necessary, or how clean the laboratory and the sample preparation environment should be. As mentioned in Chapter 15 on "Contamination," this is a major consideration if ICP-MS is the technique of choice.

User

This will tell you the required skill level of the operator; how easy the instrument is to use or what training is required. The expertise of the operator should not be underestimated if ICP-MS is being seriously considered, because it will generally require an analyst with a higher skill level to develop good methodology.

Financial

Financial factors must be considered because the funds available might have to cover the cost of the instrumentation, a specialized laboratory and/or the salary of a dedicated expert to run the instrument. Sometimes financial aspects can be the dominant reason why a technique is purchased and certainly has been a big factor in the relatively slow acceptance of ICP-MS.

Compare the Techniques

Going through these basic steps could possibly have narrowed the field to one technique or another. At this point, it may become clear that ICP-MS is the right technique. However, if this is not the case, and there is still more than one candidate technique, a detailed comparison should now be made to make the final selection. The following criteria should be used as a guideline to help in this final selection process:

- Detection limits
- Analytical working range
- Sample throughput
- Interferences
- Usability issues
- Cost of ownership

Detection Limits

The detection limits achievable for individual elements represent a significant criterion of the selection of an analytical technique for a given application problem. Without adequate detection limit capabilities, lengthy analyte concentration procedures may be required prior to analysis. Typical detection limit ranges for the major atomic spectroscopy techniques are shown in Figure 19.2.

There is no question that the best detection limits are obtained using ICP-MS followed closely by graphite furnace AA (ETA). Axial ICP-OES offers very good detection limits for most elements, but generally not as low as ETA. Radial ICP-OES and FAA show approximately the same detection limits performance, except for the refractory and the rare earth elements, which are much better by ICP-OES, because it is very difficult to produce enough ground-state atoms by FAA. For mercury and those elements that form volatile hydrides, such as As, Bi, Sb, Se, and Te, the cold vapor or hydride generation techniques offer exceptional detection limits. It is also worth mentioning that the detection capability of ICP-MS is continually being improved. Used in conjunction with collision/reaction cell or magnetic sector technology, ICP-MS is now capable of low parts-per-quadrillion (ppq) detection limits, for many elements.

Analytical Working Range

The analytical working range can be considered the concentration range over which quantitative results can be obtained without having to recalibrate the instrument. Selecting a technique with an analytical working range (and detection limits) based on the expected analyte concentrations minimizes analysis times, by allowing samples with varying analyte concentrations to be analyzed together. For example, ICP-MS, once considered exclusively an ultratrace element technique, can now handle concentration ranges from low parts-per-trillion (ppt) level up to high parts per million (ppm). A wide ana-

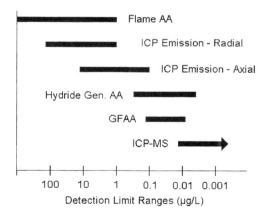

FIGURE 19.2 Typical detection limit ranges for the major atomic spectroscopy techniques.

lytical working range also can reduce sample-handling requirements and minimize potential errors. It should also be emphasized that although the dynamic range of radial and axial ICP-OES is the same, the working range of an axial ICP is shifted down approximately an order of magnitude, because the detection limits are 2–10× lower. However, there are combination systems on the market that offer both the benefits of radial and axial viewing. Figure 19.3 shows typical analytical working ranges.

Sample Throughput

Sample throughput is the number of samples that can be analyzed (or elements determined) per unit time. For most techniques, analyses performed at the limit of detection or where the best precision is required will be more time consuming than less demanding analyses. Where this is not the limiting factor, the number of elements to be determined per sample and the analytical technique will determine the sample throughput. Let us take a brief look at the sample throughput capability of each technique.

Flame AA. Flame AA provides exceptional sample throughput when analyzing a large number of samples for just a few elements. A typical determination of a single element requires only 5–10 sec. However, FAA requires specific light sources and optical parameters for each element to be determined and may require different flame gases for different elements. In automated multielement FAA systems, all samples are usually analyzed for one element, the system then automatically changes conditions for the next element, and so on until all the elements have been determined. As a result,

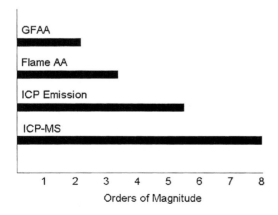

FIGURE 19.3 Analytical working ranges for the major atomic spectroscopy techniques.

although it has been used for multielement analysis, FAA is generally considered to be a single-element technique.

It should be pointed out that there are now FAA instruments on the market that are achieving higher sample throughput by carrying out "element sequential" analysis. Most traditional AA instrumentation is operated in "sample sequential" mode, where every sample in an autosampler run is analyzed for one element at a time, until all the elements in the multielement run are determined. However, by using the instrument in "element sequential" mode all the elements are determined one sample at a time by changing operating conditions such as lamp selection, slit width, wavelength, and gas flows, until all the samples in the autosampler are analyzed. The benefit of this approach is that it minimizes the time spent aspirating and flushing the sample through the tubing, nebulizer, and spray chamber—which translates into higher sample throughput and lower gas consumption. In fact, manufacturers of this technology make claims of at least a 25% improvement in productivity over traditional FAA instruments.

Electrothermal Atomization. As with FAA, ETA is basically a single-element technique, although multielement instrumentation is available from some vendors. Because of the need to thermally and sometimes chemically pretreat the sample to remove solvent and matrix components prior to atomization, ETA has a relatively low sample throughput. A typical graphite furnace determination normally requires 2–3 min per element per replicate, although multielement systems are capable of achieving up to six elements in the same amount of time.

ICP-OES. ICP-OES is commercially available as either a scanning instrument (elements determined sequentially) or a fixed channel instrument (elements determined simultaneously). The simultaneous design is usually faster, but both systems offer exceptional sample throughput capability and can determine up to 20–30 elements in a few minutes. However, when only a few elements are required, ICP probably is not the best technique, because of the relatively long read delay times of 60–90 sec, in order to wash-out/wash-in a sample and wait for the signal to reach equilibrium.

ICP-MS. ICP-MS is also a rapid multielement technique. The sample throughput of a quadrupole-based ICP-MS, which represents the majority of instruments being used for routine applications, is similar to a simultaneous ICP-OES system and is typically 20–30 elemental determinations in a few minutes, depending on such factors as the concentration levels and precision required.

Because of the many variations in sample workload and elemental requirements of different labs, it is very difficult to make a direct sample throughput comparison between the techniques. However, Table 19.1 gives a general guideline of the sample throughput capabilities of the four major AS techniques, based on the number of samples that can be analyzed per hour. It should be emphasized that these data are not absolute and should be used for comparison purposes only, but they clearly show if many analytes are being determined; ICP-OES or ICP-MS is the preferred technique. Also keep in mind that this table does not reflect the detection limits of each technique, so although ICP-OES might be as fast as ICP-MS, it obviously does not have the same kind of detection capability.

Interferences

Few, if any, of the most common analytical techniques are free of interferences. However, with atomic spectroscopy techniques, most of the common interferences have been studied and documented. As a result, methods exist to correct or compensate for those interferences. A summary of the most common interferences seen in atomic spectroscopy, and the corresponding methods of compensation, is shown in Table 19.2.

Usability

It is often said that the strength of any technique is the time it takes to set-up methods and run routine samples. The three criteria that impact a technique's ability to be considered truly routine are ease-of-use, the skill level of the operator, and whether application methodology is readily available. Here is a brief comparison of the four techniques with regard to usability.

Flame AA. Flame AA is very easy to use. It is now considered truly routine and requires minimal operator skill level. Extensive applications in-

TABLE 19.1 Comparison of Sample Throughput of the Four AS Techniques

Technique	Elements at a time	Duplicate analysis (min)	Samples per hour (1 element)	Samples per hour (5 elements)	Samples per hour (20 elements)
FAA	1	0.3	180	36	9
ETA (single)	1	5	12	2–3	1
ETA (multi)	2–6	5	12	12	3
ICP-OES	Up to 70	3	20	20	20
ICP-MS	Up to 70	3	20	20	20

TABLE 19.2 Common Types of Interferences Seen in Atomic Spectroscopy

Technique	Type of interference	Method of compensation
FAA	Ionization	Ionization buffers
	Chemical	Releasing agents or nitrous oxide-acetylene flame
	Physical	Dilution, matrix matching, or method of additions
ETA	Physical, chemical	Standard temperature platform furnace (STPF) conditions, matrix modifiers, standard additions
	Molecular absorption	Zeeman or continuum source background correction
	Spectral	Zeeman background correction
ICP-OES	Spectral	Background correction or the use of alternate analytical lines
	Matrix	Internal standardization
ICP-MS	Spectral	Interelement correction (mathematical equations); use of alternate masses; cool plasma, higher resolution systems; reaction/collision cell technology
	Matrix (physical and space charge)	Internal standardization; ion lens optimization

formation is available. Excellent precision makes it a preferred technique for the determination of major constituents and higher concentration analytes.

Electrothermal Atomization. Graphite furnace applications are well documented, although not as complete as FAA. It has exceptional detection limit capabilities but with a limited analytical working range. Sample throughput is less than that of other atomic spectroscopy techniques. Operator skill requirements are much more extensive than for FAA.

ICP-OES. This is the most widely used multielement atomic spectroscopy technique, with excellent sample throughput and very wide analytical range. Operator skill requirements are somewhere between FAA and ETA. ICP-OES is now a mature technique, which means that good applications literature is available.

ICP-MS. ICP-MS is a relatively new technique compared to the others. It has exceptional multielement capabilities at trace and ultratrace levels and also has the unique ability to perform isotopic analyses. Application information is not as readily available as the other techniques but is growing rapidly. However, ICP-MS probably requires operators with a higher skill level to achieve good quality data.

Cost of Ownership

The initial purchasing cost is obviously a big factor on the cost of ownership, but also the running costs, and the cost of consumables and chemicals will also have a big impact, particularly over the 10-year lifetime of owning the instrument. Let us first take a look at the typical purchase price of each technique. There is no question that single-element techniques (FAA and ETA) are generally less expensive than the multielement ones (ICP emission and ICP-MS). There can also be a considerable variation in cost among instrumentation of the same technique. Instruments offering only basic features are generally less expensive than more versatile systems, which frequently also offer a greater degree of automation. Figure 19.4 provides a comparison of typical cost ranges for the major atomic spectroscopy techniques. As a rough guideline, the scale starts at about $10–30K for FAA, $25–50K for ETA, $60–100K for ICP-OES, $130–200K for quadru-pole (collision/reaction cell instruments will be at the higher end of this range) or TOF ICP-MS, and about $250K and above for top-of-the-range magnetic sector systems (prices will also vary based on different geographical regions of the world).

Let us now take a look at the cost of running each of the techniques. The initial purchase price is important, but the operating costs, the price of consumables, and chemicals/standards should have a much bigger impact on the decision as to which technique to invest in—because most labs typically keep an instrument for 8–10 years before they replace it. So when calculating the overall cost of owning an instrument, it is absolutely essential that this is

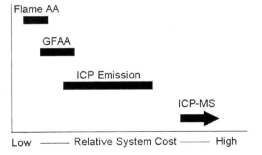

FIGURE 19.4   Relative purchasing costs of different AS equipment.

factored into the decision. So to help decide whether you can actually afford to run and operate an ICP-MS, here is a basic comparison between the running costs of the major AS techniques (2).

For the purpose of this study, let us make the assumption that the major operating costs associated with running AS instrumentation are the gases, electricity, and consumable supplies. Although the salary of the operator, lab space, and sample preparation can legitimately be called an operating expense, they will not be used for this exercise. For comparison purposes, the evaluation will be based on a typical lab running their instrument 2 ½ days (20 hr) per week and 50 weeks a year (1000 hr/year). Let us take a closer look at the cost of gases, electricity, and consumable supplies.

Gases

Flame AA. Most FAA systems use acetylene (C_2H_2) as the combustion gas and air or nitrous oxide (N_2O) as the oxidant. Air is usually generated by an air compressor, but the C_2H_2 and N_2O come in high-pressure cylinders. Normal atomic absorption grade C_2H_2 cylinders contain about 350 ft^3 (10,000 L) of gas. N_2O is purchased by weight and comes in cylinders containing about 50–60 lb of gas, which is equivalent to about 480 ft^3 (14,000 L). They both cost $200 a cylinder. Normal C_2H_2 gas flows in FAA are typically 2 L/min. At this flow rate, a cylinder will last about 80 hr.

Air/C_2H_2 is the most common gas mixture to use, while N_2O/ C_2H_2 has traditionally been used for the more "refractory" elements. For this costing exercise, we will assume that no N_2O is being used. Therefore based on a typical laboratory running the instrument for 1000 hr/year, it will consume 12 cylinders of C_2H_2, which is equivalent to $2500 per year. (Note: If N_2O elements are being determined, a cylinder will last 25–30 hr).

Electrothermal Atomization. The only gas that electrothermal atomization uses on a routine basis is high purity argon, which costs about $100 for a 350-ft^3 (10,000 L) cylinder. Typically, argon gas flows of up to 300 mL/min are required to keep an inert atmosphere in the graphite tube. At these flow rates, 500 hr can be expected out of one cylinder. Therefore a typical lab running their instrument for 1000 hr/year would consume two cylinders at a cost of about $200.

ICP-OES and ICP-MS. The consumption of gases in ICP-OES and ICP-MS is very similar. They both consume about 15–20 L/min of gaseous argon, which means a cylinder of argon would only last about 10 hr. For this reason, most users install a liquid supply of argon. A typical tank of liquid argon contains about 180 L of gas, which is equivalent to 4800 ft^3 (136,000 L) of gaseous argon and costs about $500. At 17 L/min total gas flow, this would last about 135 hr. Again, assuming a typical lab runs their instrument for

1000 hr/year, this translates to seven fills at $500 each, which is equivalent to about $3500 per year. The use of cylinders would elevate this cost to over $10,000 per year. (Note: Liquid argon will naturally bleed off and be lost to the atmosphere. For this reason, argon cylinders are probably the best option for labs that use their instruments infrequently.)

Electricity

Calculations for power consumption are based on electricity costing about 20 cents per kilowatt/hour (kW/hr). This will vary depending on the location and demand but represents a good approximation for this exercise.

Flame AA. The power in a FAA system is basically used for the hollow cathode lamps and the on-board microprocessor, which controls functions such as burner head position, lamp selection, photo multiplier tube voltage, and grating position, etc. A typical instrument requires <1000 W of power. If it is used for 1000 hr/year, it will be drawing less than 1000 kW total power, which is about $200 per year.

Electrothermal Atomization. A graphite furnace system uses considerably more power than a FAA, because a separate power unit is used to heat the graphite tube. In routine operation, there is a slow ramp heating of the tube for ~3 min until it reaches an atomization temperature of 2700°C. At this temperature, a maximum power of ~3.5 kW is required for 10–20 sec. This heating cycle combined with the power requirements for the rest of the instrument represents a cost of ~$400, for a system that is run 1000 hr/year.

ICP-OES and ICP-MS. Both these techniques can be considered the same with regard to power requirements, because the RF generators are of very similar design. Based on the voltage, magnitude of the current and the number of lines used, the majority of modern instruments draw about 4–5 kW of total power. This works out to be about $1000 for an instrument that is run 1000 hr/year.

Consumables

Because of the inherent differences between the major AS techniques, it is important to understand that there are considerable differences in the cost of consumables.

Flame AA. The major consumable supplies used in FAA are the hollow cathode lamps. Depending on usage, you should plan to replace 2–3 of them every year, at a cost of $300–400 for a good-quality, single-element lamp. Another minor cost is nebulizer tubing and autosampler tubes. These are relatively inexpensive but should be planned for. Lamps, nebulizer tubing, and a sufficient supply of autosampler tubes should not exceed more than $1300 per year, based on 1000 hr of instrument usage.

Electrothermal Atomization. As long as the sample type is not too corrosive, a GFAA tube should last about 300 heating cycles (firings). Based on a normal heating program of 3 min per replicate, this represents 20 firings per hour. If the lab is running the instrument 1000 hr/year, it will carry out a total of 20,000 firings and use 70 graphite tubes in the process. There are many designs of graphite tubes, but for this exercise, we will base the calculation on using platform-based tubes, which cost about $50 each. If we add the cost of graphite contact cylinders, hollow cathode lamps, and a sufficient supply of autosampler cups, the total cost of consumables for a graphite furnace will be approximately $5000 per year.

ICP-OES. The main consumable supply in ICP-OES is the torch itself, which consists of three concentric quartz tubes. There are many different designs available, but they all cost about $500 for a complete system. Depending on sample workload and matrices being analyzed, it is normal to go through a torch every 4–6 months. When o-rings, RF coil, spare nebulizer components, peristaltic pump tubing, and autosampler tubes are added to this, the annual cost of consumables for ICP-OES is about $2300. [Note: for this exercise, we will not include the cost of a power amplifier (PA) tube, which had a lifetime of approximately 1 year, in older style RF generators.]

ICP-MS. Besides the plasma torch and sample introduction supplies, ICP-MS requires consumables that are situated inside the mass spectrometer. The first area is the interface region between the plasma and the mass spectrometer, which contains the sampler and skimmer cones. These are traditionally made of nickel, which is recommended for most matrices, or platinum, for highly corrosive samples and organics matrices. A set of nickel cones cost about $1000, while a set of platinum cones are about $3000. Two sets of nickel cones and perhaps one set of platinum cones would be required per year. The other major consumable in ICP-MS is the detector, which has a lifetime of approximately 1 year, at a cost of about $2500. Some systems also have a replaceable ion lens. It is suggested that five of these, at $100 each, are required for a routine lab. When all these are added together with gaskets, vacuum pump consumables, investing in ICP-MS supplies represents an annual cost of about $11,000.

It should be noted that calibration standards, reference materials, chemicals, solutions, and acids are also something you have to plan for but will not be used in this evaluation, because they are not really considered instrument running costs. However, to carry out a complete assessment of each of the four techniques, they should be factored in. For example, in ICP-MS, multielement standards are generally less expensive than purchasing the same number of single-element standards. In FAA, it is fairly common to use ionization buffers to minimize the effects of easily ionizable elements. In ETA, matrix modifiers are widely used to change the volatility of analyte or matrix elements. While in ICP-OES and ICP-MS, internal standards are used in the majority of analyses, especially if the sample matrices are different from the calibration standards. In addition, sample preparation can be far more complex for single-element techniques, because it might take more than one dissolution and/or dilution step, to determine all the analytes in a multiele-ment suite. On the other hand, the cost of investing in all the necessary clean room equipment is going to be far more expensive for ICP-MS than with any other technique (with the exception maybe of ETA). And finally, if one of the techniques is going to be used with a dedicated solid sampling accessory such as a laser ablation device, the fact that no acids, chemicals/solutions, or dilutions are required will significantly impact the overall cost of analysis—both in time saving and cost of materials.

Although these sample preparation-based operating costs will not be included in this exercise, we can approximate the annual cost of gases, power, and consumable supplies of the four AS techniques—as shown in Table 19.3.

Cost per Sample

We can take this a step further and use these numbers to calculate the op-erating costs per individual sample assuming that a laboratory is determin-ing 10 analytes per sample. Let us now take a look at each technique to see how many samples can be analyzed, assuming the instrument runs 1000 hr/year.

Flame AA. A duplicate analysis for a single analyte in FAA takes about 20 sec. This is equivalent to 180 analytes/hr or 180,000 analytes/year. For 10 analytes, this represents 18,000 samples/year. Based on an annual operating cost of $4000, this equates to $0.22 per sample.

Electrothermal Atomization. A single analyte by ETA takes about 5–6 min for a duplicate analysis, which is equivalent to approx. 10 analytes/hr or 10,000 analytes/year. For 10 analytes/sample, this represents 1000 sam-ples/year. Based on an annual operating cost of $5600, this equates to $5.60

TABLE 19.3 Annual Instrument Operating Cost (US$) for a Lab Running an Instrument 1000 hr/year

Technique	Gases	Power	Supplies	Total
FAA	2500	200	1300	4000
ETA	200	400	5000	5600
ICP-OES	3500	1000	2300	6800
ICP-MS	3500	1000	11,000	15,500

per sample. (Note: If a multielement GFAA is being used, these costs will be reduced to $1.40 per sample, based on four elements being determined simultaneously.)

ICP-OES. A duplicate ICP-OES analysis for as many analytes as you require takes about 3 min. So for 10 analytes, this is equivalent to 20 samples/hr or 20,000 samples/year. Based on an annual operating cost of $6800, this equates to $0.34 per sample.

ICP-MS. ICP-MS also takes about 3 min to carry out a duplicate analysis for 10 analytes, which is equivalent to 20,000 samples/year. Based on an annual operating cost of $15,500, this equates to $0.78 per sample.

Operating costs for the determination of 10 analytes/sample are summarized in Table 19.4.

For labs with an extremely high sample workload requiring in excess of 20 analytes/sample, a single-element technique such as ETA becomes less of a practical option, as well as being cost-prohibitive compared to ICP-MS. Whereas the running costs of FAA are still very competitive with the multielement techniques, it is impractical in a high workload environment. On the other hand, when the elemental requirements are less demanding, FAA and ETA will look much more attractive if the running costs are based on cost per analyte. For example, for a lab that is running a set of samples that require just one analyte, the cost/sample for FAA and ETA will be $0.02 and $0.56, respectively, while the costs for ICP-OES and ICP-MS will basically remain the same. This can be seen in Table 19.5.

It must also be emphasized that this comparison does not take into account the detection limit requirements but is based on instrument operating costs alone. These figures have been generated for a "typical" workload using what would be considered "average" cost of gases, power, and consumables. Every lab's situation is unique, especially outside the United States, so for that reason this costing exercise should be treated with caution

TABLE 19.4 Operating Costs for a Sample Requiring 10 Analytes, Based on the Instrument Being Used 1000 hr/year

Technique	Operating cost/sample ($)
FAA	0.22
ETA	5.60
ICP-OES	0.34
ICP-MS	0.78

TABLE 19.5 Running Costs for a Sample
Requiring One Analyte

Technique	Operating cost/analyte ($)
FAA	0.02
ETA	0.56
ICP-OES	0.34
ICP-MS	0.78

and only used as a guideline for comparison purposes. If required, it can be taken a step further, by also including the purchase price of the instrument, the cost of installing a clean room, the cost of sample preparation, and the salary of the operator. This would be a very useful exercise, as it would give a good approximation of the overall cost of analysis and therefore could be used as a guideline for calculating what a lab might charge for running samples on a commercial basis.

CONCLUSION

It is important to remember that there are many criteria to consider when selecting a trace element technique. You have to decide which are the most important ones for your application and your laboratory. This is not meant to be an exhaustive comparison of all elemental techniques. It should be used as a guideline to evaluate the most commonly used trace and ultratrace, atomic spectroscopy-based techniques. It has been done in a very simplistic way and has not attempted to compare the many variations, features, and sampling accessories offered by the different manufacturers. However, it is clear that there is no single technique suitable for all applications. They all have their own strengths and weaknesses. It is therefore important when making the comparison that all these avenues are explored. Maybe ICP-MS is a technique that you would really like to have in your laboratory. True, it is a very powerful piece of equipment, but at the end of the day, can the purchase be really justified? In most cases I believe it can be, but it is definitely worth investing the time and effort to collect the evidence, in order to support that justification. Hopefully, this chapter has given you some insight into this process.

FURTHER READING

1. Thomas RJ. Today's Chem Work 1999; 8(10):42–48.
2. Thomas RJ. Today's Chem Work 2000; 9(9):19–25.

20

How to Select an ICP–Mass Spectrometer: Some Important Analytical Considerations

Understanding the basic principles of inductively coupled plasma mass spectrometry (ICP-MS) is important, but not absolutely essential, in order to operate and to use an instrument on a routine basis. However, understanding how these basic principles affect the performance of an instrument is a real benefit to evaluate the analytical capabilities of the technique. There is no question that the better informed you are while going into an evaluation of commercial instrumentation, the better chance you have of selecting the right one for your application. Having been involved in demonstrating ICP-MS equipment for over 10 years, I know the mistakes that people make when they get into the selection process. So in Chap. 20, I will attempt to present a set of evaluation guidelines to help you make the right decision.

So, you have convinced your boss that inductively coupled plasma mass spectrometry (ICP-MS) is perfect for your laboratory. Hopefully, the chapters on fundamental principles have given you the basic knowledge and a good platform on which to go out and evaluate the marketplace. However, they do not really give you an insight on how to compare instrument designs, hardware components, and software features, which are of critical importance when you have to make a decision as to which instrument to purchase. There are a number of commercial systems available in the marketplace, which look very similar and have very similar specifications, but how do you know which is the best one that fits your needs? This section will attempt to present a set of evaluation guidelines to help you decide on the most important figures of merit for your application. However, to get the most out of this chapter, it should be used in conjunction with other chapters of this book.

EVALUATION OBJECTIVES

Before you begin the selection process, it is very important to decide on what your analytical objectives are. This is particularly important if you are part of an evaluation committee. It is alright to have more than one objective, but it is essential that all members of the group begin the evaluation process with the objectives clearly defined. For example, is detection limit performance an important objective for your application, or is it more important to have an instrument that is easy to use? If the instrument is being used on a routine basis, maybe good reliability is also very critical. On the other hand, if the instrument is being used to generate revenue, perhaps sample throughput and cost of analysis is of greater importance. Every laboratory's scenario is unique, so it is important to prioritize before you begin the evaluation process. So as well as looking at instrument features and components, the comparison should also be made with your analytical objectives in mind. Let us take a look at the most common ones that are used in the selection process. They typically include:

> Analytical performance
> Usability aspects
> Reliability issues
> Financial considerations

Let us examine these in greater detail.

ANALYTICAL PERFORMANCE

Analytical performance can mean different things to different people. The major reason that the trace element community was attracted to ICP-MS almost 20 years ago was because of its extremely low multielement detection limits. Other multielement techniques such as inductively coupled plasma optical emission spectroscopy (ICP-OES) offered very high throughput, but just could not get down to ultra-trace levels. Even though electrothermal atomization (ETA) offered much better detection capability than ICP-OES, it did not offer the sample throughput capability that many applications demanded. In addition, ETA was predominantly a single-element technique, thus was impractical for carrying out rapid multielement analysis. These limitations quickly led to the commercialization and acceptance of ICP-MS as a tool for rapid ultra-trace element analysis. However, there are certain areas where ICP-MS is known to have weaknesses. For example, dissolved solids for most sample matrices must be kept below 0.2%; otherwise, it can lead to serious drift problems and/or poor precision. Polyatomic and isobaric inter-ferences, even in simple acid matrices, can produce unexpected spectral overlaps, which will have deleterious impact on your data. Moreover, depending on the sample being analyzed, matrix components can dramatically suppress analyte sensitivity and affect accuracy. These potential problems can all be reduced to a certain extent, but different instruments approach and compensate for these problem areas in different ways. With a novice, it is often ignorance or a basic lack of understanding of how a particular instrument works that makes the selection process more complicated that it really should be. So any information that can help you prepare for the evaluation will put you in a much stronger position.

It should be emphasized that these evaluation guidelines are based on my personal experience and should be used in conjunction with other materials in the open literature that have presented broad guidelines to compare figures of merit for commercial instrumentation [1–3]. In addition, you should talk with colleagues in the same industry or application segment as yourself. If they have gone through a lengthy evaluation process, they can give you valuable pointers, or even suggest the instrument that is better suited to your needs. Finally, before we begin, it is strongly suggested that you narrow the actual evaluation to two or maybe three commercial products. By carrying out some preevaluation research, you will have a better understanding as to what ICP-MS technology or instrument to focus on. For example, if funds are limited and you are purchasing ICP-MS for the very first time to carry out high-throughput environmental testing, it is probably more cost-effective to focus on quadrupole technology. On the other hand, if you are investing in a second system to enhance the capabilities of your quadrupole instrument, it might be worth taking a look at collision/reaction cell or magnetical sector technology. Or if fast multielement transient peak analysis is your major reason for investing in ICP-MS, turnover frequency (TOF) technology should be given serious consideration. One final note I would like to add, although it is not strictly a technical issue, is that if you are prepared to forego an instrument demonstration, or do not need any samples run, you will be in a much stronger position with an instrument vendor to negotiate a lower price. You should keep that in mind before you decide to get involved in a lengthy selection process.

So let us begin by looking at the most important aspects of instrument performance. Depending on the application, the major performance issues that need to be addressed include:

Detection capability
Precision/signal stability
Accuracy
Dynamic range
Interference reduction

Sample throughput
Transient signal capability.

DETECTION CAPABILITY

Detection capability is a term used to assess the overall detection performance of an ICP mass spectrometer. There are a number of different ways of looking at detection capability, including instrument detection limit (IDL), elemental sensitivity, background signal, and background equivalent concentration (BEC). Of these four criteria, the IDL is generally thought to be the most accurate way of assessing instrument detection capability. It is often referred to as signal-to-background noise and, for a 99% confidence level, is typically defined as $3\times$ standard deviation (SD) of n replicates ($n = \sim 10$) of the sample blank and is calculated in the following manner:

$$\text{IDL} = \frac{3 \times \text{SD of Background Signal}}{\text{Analyte Intensity} - \text{Background Signal}} \times \text{Analyte Concentration}$$

However, there are slight variations of both the definition and calculation of instrument detection limits, so it is important to understand how different manufacturers quote their DLs if a comparison is to be made. They are usually run in single-element mode, using extremely long integration times (5–10 sec), in order to achieve the highest quality data. So when comparing detection limits of different instruments, it is important to know the measurement protocol used.

A more realistic way of calculating analyte detection limit performance in your sample matrices is to use method detection limit (MDL). The MDL is broadly defined as the minimum concentration of analyte that can be determined from zero with a 99% confidence. MDLs are calculated in a manner similar to IDLs, except that the test solution is taken through the entire sample preparation procedure before the analyte concentration is measured multiple times. This difference between MDL and IDL is exemplified in EPA Method 200.8, where a sample solution at 2–5× estimated IDL is taken through all the preparation steps and analyzed. The MDL is then calculated in the following manner:

$$\text{MDL} = tS$$

where t is the Student's t value for a 95% confidence level and specifies a standard deviation estimate with $n - 1$ degrees of freedom ($t = 3.14$ for seven replicates) and S is the standard deviation of the replicate analyses.

Both IDL and MDL are very useful in understanding the capability of ICP-MS, but whatever method is used to compare detection limits of different manufacturers' instrumentation, it is essential to carry out the tests using realistic measurement times that reflect your analytical situation. For example, if you are determining a group of elements across a mass range in a digested rock sample, it is important to know how much the sample matrix suppresses the analyte sensitivity because the detection limit of each analyte will be impacted by the amount of suppression across the mass range. On the other hand, if you are carrying out high-throughput multielement analysis of drinking or wastewater samples, you probably need to be using relatively short integration times (1–2 sec per analyte) to achieve the desired sample throughput. Or if you are dealing with a laser ablation or flow injection transient peak that lasts for 10–20 sec, it is absolutely critical that you understand the impact time has on detection limits compared to a continuous signal generated with a conventional nebulizer. (In fact, analyses times and detection limits are very closely related to each other and will be discussed later on in this chapter.) In other words, when comparing instrument detection limits, it is absolutely critical that the tests represent your real-world analytical situation.

Elemental sensitivity is also a useful assessment of instrument performance, but it should be viewed with caution. It is usually a measurement of background corrected intensity at a defined mass and is typically specified as counts per second (cps) per concentration [parts per billion (ppb) or parts per million (ppm)] of a midmass element such as $^{103}Rh^+$ or $^{115}In^+$. However, unlike detection limit, raw intensity usually does not tell you anything about the intensity of the background, or the level of the background noise. It should be emphasized that instrument sensitivity can be enhanced by optimization of operating parameters such as radiofrequency (RF) power, nebulizer gas flows, torch sampling position, interface pressure, and sampler/skimmer cone geometry, but usually comes at the sacrifice of other performance criteria, including oxide levels, matrix tolerance, or background intensity. So be very cautious when you see an extremely high sensitivity specification because there is a strong probability that the oxide or background specifications might also be high. For that reason, it is unlikely there will be an improvement in detection limit unless the increase in sensitivity comes with no compromise in the background level. It is also important to understand the difference between background and background noise when comparing specifications (background noise is a measure of the stability of the background and is defined as the square root of the background signal). Most modern quadrupole instruments today specify 20–50 million cps/ppm rhodium ($^{103}Rh^+$) or indium ($^{115}In^+$) and <10 cps of background (usually at 220 amu), whereas

magnetical sector instrument sensitivity specifications are typically 10–20×
higher, with 10× lower background.

Another figure of merit that is being used more routinely nowadays is
background equivalent concentration. BEC is defined as the intensity of the
background at analyte mass, expressed as an apparent concentration and is
typically calculated in the following manner:

$$\text{BEC} = \frac{\text{Intensity of Background Signal}}{\text{Analyte Intensity} - \text{Background Intensity}} \times \text{Analyte Concentration}$$

It is considered more of a realistic assessment of instrument performance in real-world sample matrices (especially if the analyte mass sits on a
high background) because it gives an indication of the level of the background—defined as a concentration value. Detection limits alone can sometimes be misleading because they are influenced by the number of readings
taken, integration time, cleanliness of the blank, and at what mass the background is measured, and are rarely achievable in a real-world situation.
Figure 20.1 emphasizes the difference between detection limit and background equivalent concentration. In this example, 1 ppb of an analyte produces a signal of 10,000 cps and a background of 1000 cps. Based on the
calculations defined earlier, BEC is equal to 0.11 ppb because it is expressing

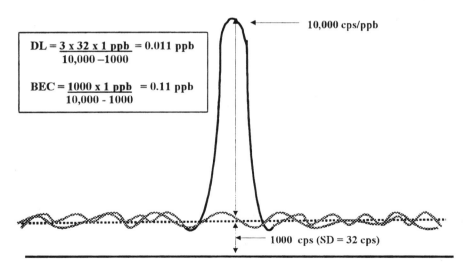

FIGURE 20.1 DL is calculated using the noise of the background, whereas BEC is
calculated using the intensity of the background.

the background intensity as a concentration value. On the other hand, DL is 10× lower because it is using the standard deviation of the background (i.e., the noise) in the calculation. For this reason, BECs are particularly useful when it comes to comparing the detection capabilities of techniques such as cool plasma and collision/reaction cell technology because it gives you a very good indication of how efficient the background reduction process is.

It is also important to remember that peak measurement protocol will also have an impact on detection capability. As mentioned in Chap. 12 on "Peak Integration," there are basically two approaches to measuring an isotopic signal in ICP-MS. There is the multichannel scanning approach, which uses a continuous smooth ramp of 1–20 channels per mass across the peak profile, and there is the peak hopping approach, where the mass analyzer power supply is driven to a discrete position on the peak, allowed to settle, and a measurement taken for a fixed period of time. This is usually at the peak maximum, but can be as many points as the operator selects. This process is simplistically shown in Figure 20.2.

The scanning approach is best for accumulating spectral and peak shape information when doing mass calibration and resolution scans. It is traditionally used as a classical method development tool to find out what elements are present in the sample and to assess spectral interferences on the masses of interest. However, when the best possible detection limits are required, it is clear that the peak hopping approach is best. It is important to understand

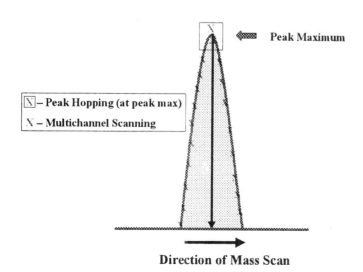

Direction of Mass Scan

FIGURE 20.2 There are typically two approaches to peak quantitation: peak hopping (usually at peak maximum; X in box) and multichannel scanning (X).

that to get the full benefit of peak hopping, the best detection limits are achieved when single-point peak hopping at the peak maximum is chosen. It is well accepted that measuring the signal at the peak maximum will always give the best signal-to-background noise for a given integration time, and that there is no benefit to spread your available integration time over more than one measurement point per mass [4]. Instruments that use more than one point per peak for quantitation are sacrificing measurement time on the sides of the peak, where the signal-to-noise is worse. However, the ability of the mass analyzer to repeatedly scan on the same mass position every time during a multielement run is of paramount importance for peak hopping. If multiple points per peak are recommended, it is a strong indication that the spectrometer has poor mass calibration stability because it cannot guarantee that it will always find the peak maximum with just one point. Mass calibration specification, which is normally defined as a shift in peak position (in atomic mass units) over an 8-hr period, is a good indication of mass stability. However, it is not always the best way to compare systems because peak algorithms using multiple points are often used to calculate the peak position. A more accurate way is to assess the short-term and long-term mass stability by looking at relative peak positions over time. Short-term stability can be

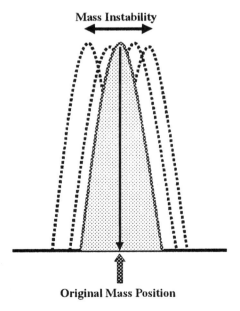

FIGURE 20.3 Good mass stability is critical for single-point peak hopping quantitation.

determined by aspirating a multielement solution containing four elements (across the mass range) and recording spectral profiles using multichannel ramp scanning of 20 points per peak. Now repeat the multielement scan 10× and record the peak position of every individual scan. Calculate the average and the relative standard deviation (RSD) of the scan positions. The long-term mass stability can then be determined by repeating the test 8 hr later to see how far the peaks have moved. It is important of course that mass calibration procedure is not carried out during this time. Figure 20.3 shows what might happen to the peak position over time if the analyzer's mass stability is poor.

PRECISION

Short-term and long-term precision specifications are usually a good indication of how stable an instrument is (refer to Chap. 12 on "Peak Measurement Protocol"). Short-term precision is typically specified as percent RSD of 10 replicates of 1–10 ppb of three elements across the mass range using 2–3 sec integration times, whereas long-term precision is a similar test, but normally carried out every 5–10 min over a 4- to 8-hr period. Typical short-term precision, assuming an instrument warm-up time of 30–40 min, should be approximately 1–3%, whereas long-term precision should be on the order of 3–5%—both determined without using internal standards. However, it should be emphasized that under these measurement protocols, it is unlikely that you will see a big difference in the performance between different instruments in simple aqueous standards. A more accurate reflection of the stability of an instrument is to carry out the tests using a typical matrix that would be run in your laboratory at the concentrations you expect. It is also important that stability should be measured without the use of an internal standard. This will enable you to evaluate the instrument drift characteristics, without any type of signal correction method being applied.

It is recognized that the major source of drift and imprecision in ICP-MS, particularly with real-world samples, is associated with either the sample introduction area, design of the interface, or the ion optics system. Some of the common problems encountered are:

Pulsations and fluctuations in the peristaltic pump, leading to increased signal noise

Blockage of the nebulizer over time resulting in signal drift, especially if the nebulizer does not have a tolerance for high dissolved solids

Poor drainage, producing pressure changes in the spray chamber and resulting in spikes in the signal

Build-up of solids in the sample injector, producing signal drift

Changes in the electrical characteristics of the plasma, generating a secondary discharge and increasing ion energies

Blockage of the sampler and skimmer cone orifice with sample material, causing instability

Erosion of the sampler and skimmer cone orifice with high concentration acids

Coating of the ion optics with matrix components, resulting in slight changes in the electrical characteristics of ion lens system.

These are all relative problems depending on the types of samples being analyzed. However, the most common and potentially serious problem with real-world matrices is the deposition of sample material on the interface cones and the ion optics over time. It does not impact short-term precision that much because careful selection of internal standards, matched to the analyte masses, can compensate for slight instability problems. However, sample materials, particularly matrix components found in environmental, clinical, and geochemical samples, can have a dramatical effect on long-term stability. The problem is exaggerated even more if you are a high-throughput laboratory because poor stability will necessitate the need for more regular recalibration and might even require some samples to be rerun if quality control (QC) standards fall outside certain limits. There is no question that if an instrument has poor drift characteristics, it will take much longer to run an autosampler tray full of samples and, in the long term, result in much higher argon consumption.

For these reasons, it is critical that when short-term and long-term precision is evaluated, you know all the potential sources of imprecision and drift. For that reason, it is important that you either choose a matrix that is representative of your samples, or you select a matrix that will genuinely test the instrument out. Typical sample matrices include:

Drinking waters, containing calcium and magnesium salts at a few hundred parts per million

Rock digests, containing calcium, magnesium, iron, and aluminum at a few hundred parts per million, with maybe some alkaline peroxide/borate fusion mixtures

Biological fluids such as blood or urine, containing carbonaceous, organic, and saline components

Saline samples, containing sodium, magnesium, and calcium chlorides

Metallurgical alloys, containing concentrations of various metals dissolved in 1–5% mineral acids

Organic samples, such as diluted oils, alcohols, ketones, or aromatic solvents.

Whatever matrices are chosen, it must be emphasized that for the stability test to be meaningful, no internal standards should be used, the sample should contain less than 0.2% total dissolved solids, and the representative elements should be at a reasonably high concentration (1–10 ppb) and be spread across the mass range. In addition, no recalibration should be carried out for the length of the test, and should reflect your real-world situation [5]. For example, if you plan to run your instrument in a high-throughput environment, you might want to carry out an 8-hr or even an overnight (12–16 hr) stability test. If you are not interested in such long runs, a 2- to 4-hr stability test will probably suffice. But just remember, plan the test beforehand and make sure you know how to evaluate the vast amount of data it will generate. It will be hard work, but I guarantee that it is worth it in order to fully understand the short-term and long-term drift characteristics of the instruments you are evaluating.

Isotope Ratio Precision

An important aspect of ICP-MS is its ability to carry out fast isotope ratio precision data. With this technique, two different isotopes of the same element are continuously measured over a fixed period of time. The signal of one isotope is ratioed to the other, and the precision of the ratios is then calculated. Analysts who are interested in isotope ratios are usually looking for the ultimate in precision. The optimum way to achieve this in order to get the best counting statistics would be to carry out the measurement simultaneously with a multicollector magnetical sector instrument or a TOF ICP-MS system. However, a quadrupole mass spectrometer is a rapid sequential system, so the two isotopes are never measured at exactly the same moment in time. This means that the measurement protocol must be optimized in order to get the best precision. As discussed earlier, the best and most efficient use of measurement time is to carry out single-point peak hopping between the two isotopes. In addition, it is also beneficial to be able to vary the total measurement time of each isotope, depending on their relative abundance. The ability to optimize the dwell time and the number of sweeps of the mass analyzer ensures that the maximum amount of time is being spent on top of each individual peak where the signal-to-noise is at its best [6].

It is also critical is to optimize the efficiency cycle of the measurement. With every sequential mass analyzer, there is an overhead time, called a settling time, to allow the power supply to settle before taking a measurement. This time is often called nonanalytical time because it does not contribute to the quality of the analytical signal. The only time that contributes to the analytical signal is the dwell time, or the time that is actually spent measuring the peak. The measurement efficiency cycle (MEC) is a ratio of the dwell time compared to the total analytical time including settling time and is expressed as:

$$MEC(\%) = \frac{Number\ of\ Sweeps \times Dwell\ Times}{\{Number\ of\ Sweeps(Dwell\ Time + Setting\ Time)\}} \times 100$$

It is therefore obvious that to get the best precision over a fixed period of time, the settling time must be kept to an absolute minimum. The dwell time and the number of sweeps are operator-selectable, but the settling time is usually fixed because it is a function of the quadrupole electronics. For this reason, it is important to know what the settling time of the mass spectrometer is when carrying out peak hopping. Remember, a shorter settling time is more desirable because it will increase the measurement efficiency cycle and improve the quality of the analytical signal [7].

In addition, if isotope ratios are being determined on vastly different concentrations of major and minor isotopes using the extended dynamic range of the system, it is important to know the settling time of the detector electronics. This settling time will affect the detector's ability to detect the analog and pulse signals (or in dynamic attenuation mode with a pulse-only EDR (extended dynamic range) system) when switching between measurements of the major and minor isotopes, which could have a serious impact on the accuracy and precision of the isotope ratio. So for that reason, no matter how the higher concentrations are handled, shorter settling times are more desirable, thus switching/attenuation can be carried out as quickly as possible.

This is exemplified in Figure 20.4, which shows a spectral scan of $^{63}Cu^+$ and $^{65}Cu^+$ using an automated pulse/analog EDR detection system. The natural abundance of these two isotopes is $^{63}Cu^+$–69.17% to $^{65}Cu^+$–30.83%. However, the ratio of these isotopes has been artificially altered to be $^{63}Cu^+$ –0.39% to $^{65}Cu^+$–99.61%. The intensity of ^{63}Cu is about 70,000 cps, whereas the intensity of the $^{65}Cu^+$ is about 10 million cps, which necessitates the need for pulse counting for the $^{63}Cu^+$ and analog counting for the $^{65}Cu^+$. There is no question that the counting circuitry would miss many of the ions and generate erroneous concentration data if the switching between pulse and analog modes is not fast enough.

So when evaluating isotopic ratio precision, it is important that the measurement protocol and peak quantitation procedure can be optimized. Quoted specifications will be a good indication as to what the instrument is capable of, but once again, these will be defined in aqueous-type standards, using relatively short total measurement times (typically 5 min). For that reason, if the test is to be meaningful, it should be optimized to reflect your real-world analytical situation.

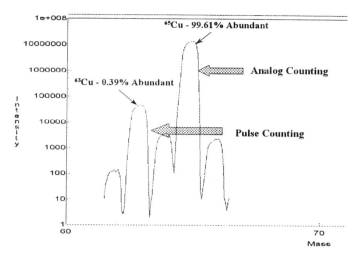

FIGURE 20.4   The detector electronics must be able to switch fast enough to detect isotope ratios that require both pulse and analog counting modes. (Courtesy of Perkin-Elmer Life and Analytical Sciences.)

ACCURACY

Accuracy is a very difficult aspect of instrument performance to evaluate because it often reflects the skill of the person developing the method and it entails analyzing the samples, instead of measuring the capabilities of the instrument itself. If handled correctly, it is a very useful exercise to go through, particularly if you can get hold of reference materials (ideally of matrices similar to your own) whose values are well defined. However, if you attempting to compare the accuracy of different instruments, it is essential that you prepare every sample yourself, including the calibration standards, blanks, unknown samples, QC standards, or certified reference materials (CRMs). I suggest that you make up enough of each solution to give to each vendor for analysis. By doing this, you eliminate the uncertainty and errors associated with different people making up different solutions. It then becomes more of an assessment of the capability of the instrument, including its sample introduction system, interface region, ion optics, mass analyzer, and detector and measurement circuitry to handle the unknown samples, minimize interferences, and get correct results.

A word of caution should be expressed at this point. Having been working with ICP-MS for almost 20 years, I know that the experience of the person developing the method, running the samples, and doing the demonstration has a direct impact on the quality of the data generated in ICP-MS. There is no question in my mind that the analyst with the most application expertise has a much better chance of getting the right answer than someone who is either inexperienced, or is not familiar with a particular type of sample. I think it is valid to compare the ability of the application specialist because this might be the person who will be giving you technical support. However, if you want to assess the capabilities of the instrument alone, it is essential to take the skill of the operator out of the equation. This is not as straightforward as it sounds, but I have found that the best way to "level the playing field" is to send some of your sample matrices to each vendor before the actual demonstration. This allows the application person to spend time developing the method and to get familiar with the samples. You can certainly hold back on your CRM or QC standards until you get to the demonstration, but at least it gives each vendor some uninterrupted time with your samples. This also allows you to spend most of the time at the demonstration evaluating the instrument, assessing hardware components, comparing features, and getting a good look at the software. It is my opinion that most instruments on the market should get the right answer—at least for the majority of routine applications. So, even though the accuracy of different instruments should be compared, it is more important to understand how the result was achieved, especially when it comes to the analysis of very difficult samples.

DYNAMIC RANGE

When ICP-MS was first commercialized, it was primarily used to determine very low analyte concentrations. As a result, detection systems were only asked to measure concentration levels up to approximately five orders of magnitude. However, as the demand for greater flexibility grew, it was called upon to extend its dynamic range in order to determine higher and higher concentrations. Today, the majority of commercial systems come standard with detectors that can measure signals up to eight orders of magnitude.

As mentioned in Chap. 11 on "Detectors," there are subtle differences between the way various detectors and detection systems achieve this, so it is important to understand how different instruments extend the dynamic range. The majority of quadrupole-based systems on the market extend the dynamic range by using a discrete dynode detector operated either in pulse-only mode, or a combination of pulse and analog mode. When evaluating this feature, it is important to know whether this is done in one or two scans because it will have an impact on the types of samples you can analyze. The different approaches have been described earlier, but it is worth briefly going through them again:

> **Two-scan approach:** Basically, two types of two-scan or prescan approaches have been used to extend the dynamic range.

In the first one, a survey or prescan is used to determine what masses are at high concentrations and what masses are at trace levels. Then the second scan actually measures the signals by switching rapidly between analog and pulse counting.

In the second two-scan approach, the detector is first run in the analog mode to measure the high signals and then rescanned in pulse-counting mode to measure the trace levels.

One-scan approach: This approach is used to measure both the high levels and trace concentrations in one simultaneous scan. This is typically achieved by measuring the ion flux as an analog signal at some midpoint on the detector. When more than a threshold number of ions are detected, the ions are processed through the analog circuitry. When fewer than a threshold number of ions are detected, the ions cascade through the rest of the detector and are measured as a pulse signal in the conventional way.

Using pulse-only mode: The most recent development in extending the dynamic range is to use the pulse-only signal. This is achieved by monitoring the ion flux at one of the first few dynodes of the detector (before extensive electron multiplication has taken place) and then attenuating the signal by applying a control voltage. Electron pulses passed by the attenuation section are then amplified to yield pulse heights that are typical in normal pulse-counting applications. Under normal circumstances, this approach only requires one scan, but if the samples are complete unknowns, dynamic attenuation might need to be carried out, where an additional premeasurement time is built into the quadrupole settling time in order to determine the optimum detector attenuation for the selected dwell times used.

The methods that use a prescan or premeasurement time work very well, but they do have certain limitations for some applications compared to the one-scan approach. Some of these include:

The additional scan/measurement time means that it will use more of the sample. Ordinarily, this will not pose a problem, but if sample volume is limited to a few hundred microliters, it might be an issue.

If concentrations of analytes are vastly different, the measurement circuitry reaction time of a prescan system might struggle to switch quickly enough between high and low concentration elements. This is not such a major problem, unless the measurement circuitry has to switch rapidly between consecutive masses in a multielement run, or there are large differences in the concentrations of two isotopes of the same element when carrying out ratio studies. In both these situations, there is a possibility that the detection system will miss counting some of the ions and produce erroneous data.

The other advantage of the one-scan approach is that more time can be spent measuring the peaks of interest in a transient peak, generated by a flow injection or laser sampling system that only lasts a few seconds. With a detector that uses two scans or a prescan approach, you can use a large amount of the available time just to characterize the sample. It is exaggerated even more with a transient peak, if the analyst has no prior knowledge of elemental concentrations in the sample.

This final point is exemplified in Figure 20.5, which shows the measurement of a flow injection peak of NIST 1643C potable water CRM, using an automated simultaneous pulse/analog EDR system. It can be seen that K and Ca are at parts-per-million levels, which requires the use of the analog counting circuitry, whereas Pb and Cd are at parts-per-billion levels, which requires pulse counting [8]. This would not be such a difficult analysis for a detector, except that the transient peak only lasted 10 sec. This means that in order to get the highest quality data, you have to spend all the available time quantifying the peak. In other words, you cannot afford the luxury of doing a premeasurement, especially if you have no prior knowledge of the analyte concentrations.

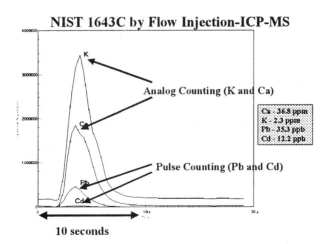

FIGURE 20.5 A one-scan approach to extending the dynamic range is more advantageous for handling a fast transient signal, such as a flow injection peak. (From Ref. 8.)

For these reasons, it is important to understand how the detector handles high concentrations in order to evaluate it correctly. If you are truly interested in using ICP-MS to determine higher concentrations, you should check out the linearity of different masses across the mass range by measuring high parts per trillion (ppt) (~500 ppt), low parts per billion (~50 ppb), and parts per million (10–100 ppm) levels. Do not be afraid to analyze a standard reference material (SRM) sample such as one of the NIST 1643 series of drinking water reference standards, which has both high (parts per million) and low (parts per billion) levels. Finally, if you know you have large concentration differences between the same analytes, make sure the detector is able to determine them with good accuracy and precision. On the other hand, if your instrument is only going to be used to carry out ultra-trace analysis, it probably is not worth spending the time to evaluate the capability of the extended dynamic range feature.

INTERFERENCE REDUCTION

As mentioned in Chap. 14 on "Interferences," there are two major types that have to be compensated for: spectral and matrix (space charge and physical) interferences. Although most instruments approach the principles of interference reduction in a similar way, the practical aspect of compensating for them will be different, based on the differences in hardware components and instrument design. Let us look at interference reduction in greater detail and compare the different approaches used.

Reducing Spectral Interferences

The majority of spectral interferences seen in ICP-MS are produced by either the sample matrix, the solvent, the plasma gas, or various combinations of them. If the interference is caused by the sample, the best approach might be to remove the matrix by some kind of ion exchange column. However, this can be cumbersome and time-consuming on a routine basis. If the interference is caused by solvent ions, simply desolvating the sample will have a positive effect on reducing the interference. For that reason, systems that come standard with chilled spray chambers to remove much of the solvent usually generate less sample-based oxide-induced, hydroxide-induced, and hydride-induced spectral interferences. There are alternative ways to reduce these types of interferences, but cooling the spray chamber can be a relatively simple way of achieving an order-of-magnitude reduction in oxide-based and hydride-based based ionic species.

Spectral interferences are an unfortunate reality in ICP-MS and it is now generally accepted that instead of trying to reduce or minimize them, the best way is to resolve the problem using high-resolution technology such as a double-focusing magnetical sector mass analyzer [9]. Even though they are not considered ideal for a routine high-throughput laboratory, they offer the ultimate in resolving power and have found a niche in applications that require ultra-trace detection and a high degree of flexibility for the analysis of complex sample matrices. If you use a quadrupole-based instrument and are looking to purchase a second system to enhance the flexibility of your laboratory, it might be worth taking a serious look at magnetical sector technology. The full benefits of this type of mass analyzer for ICP-MS have been described in Chap. 8.

Let us now turn our attention to the different approaches used to reduce spectral interferences using quadrupole-based technology. Each approach should be evaluated based on its suitability for the demands of your particular application.

Improvement of Resolution

As described in Chap. 7 on "Quadrupole Mass Analyzers," there are two very important performance specifications of a quadrupole—resolution and abundance sensitivity [10]. Although they both define the ability of a quadrupole to separate an analyte peak from a spectral interference, they are measured differently. Resolution reflects the shape of the peak and is normally defined as the width of a peak at 10% of its height. Most instruments on the market have similar resolution specifications of 0.3–3.0 amu and typically use a nominal setting of 0.7–1.0 amu for all masses in a multielement run. For this reason, it is unlikely that you will see any measurable difference when you make your comparison. However, some systems allow you to change resolution settings on the fly, on individual masses during a multielement analysis. Under normal analytical scenarios, this is rarely required, but at times, it can be advantageous to improve the resolution for an analyte mass, particularly if it is close to a large interference and there is no other mass or isotope available for quantitation. This can be seen in Figure 20.6, which shows a spectral scan of 10 ppb $^{55}Mn^+$, which is monoisotopic, and 100 ppm of $^{56}Fe^+$. The left-hand plot shows the scan using a resolution setting of 0.8 amu for both $^{55}Mn^+$ and $^{56}Fe^+$, whereas the right-hand plot shows the same scan, but using a resolution setting of 0.3 amu for $^{55}Mn^+$ and 0.8 amu for $^{56}Fe^+$. Even though the $^{55}Mn^+$ peak intensity is about 3× lower at 0.3 amu resolution, the background from the tail of the large $^{56}Fe^+$ is about 7× less, which translates into an improvement in the $^{55}Mn^+$ detection limit at a resolution of 0.3 amu, compared to 0.8 amu.

Higher Abundance Sensitivity Specifications

The second important specification of a mass analyzer is abundance sensitivity, which is a reflection of the width of a peak at its base. It is defined as

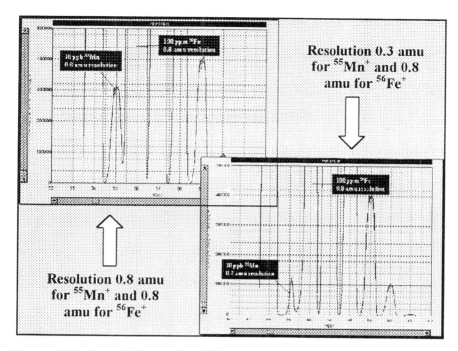

FIGURE 20.6 A resolution setting of 0.3 amu will improve the detection limit for $^{55}Mn^+$ in the presence of high concentrations of $^{56}Fe^+$. (Courtesy of Perkin-Elmer Life and Analytical Sciences.)

the signal contribution of the tail of a peak at one mass lower and one mass higher than the analyte peak and, generally speaking, the lower the specification is, the better is the performance of the mass analyzer. The abundance sensitivity of a quadrupole is determined by a combination of factors including the shape, diameter, and length of the rods; frequency of quadrupole power supply; and the slope of the applied RF/direct current (DC) voltages. Even though there are differences between designs of quadrupoles in commercial ICP-MS systems, there appears to be very little difference in their performance.

When comparing abundance sensitivity, it is important to understand what the numbers mean. The trajectory of an ion through the analyzer means that the shape of the peak at one mass lower ($M-1$) is slightly different from the other side of the peak at one mass higher ($M+1$) than the mass M. For this reason, the abundance sensitivity specification for all quadrupoles is always worse on the low mass side than the high mass side, and is typically 1×10^{-6} at $M-1$ and 1×10^{-7} at $M+1$. In other words, an interfering peak of 1 million cps at $M-1$ would produce a background of 1 cps at M, whereas it would take an interference of 10 million cps at $M+1$ to produce a background of 1 cps at M. In theory, hyperbolical rods will demonstrate better abundance sensitivity than round ones, as will a quadrupole with longer rods and a power supply with higher frequency. However, you have to evaluate whether this produces any tangible benefits when it comes to the analysis of your real-world samples.

Use of Cool Plasma Technology

Most of the instruments on the market can be set up to operate under cool or cold plasma conditions (some better than others) in order to achieve very low detection limits for elements such as K, Ca, and Fe. Cool plasma conditions are achieved when the temperature of the plasma is cooled sufficiently low enough to reduce the formation of both argon-based and solvent-based argon polyatomic species [11], as shown in Figure 20.7. It can be seen at the right-hand spectral display that the intensity of the argon-based species under cool plasma conditions is significantly less than under normal plasma conditions, shown in the left-hand spectral display.

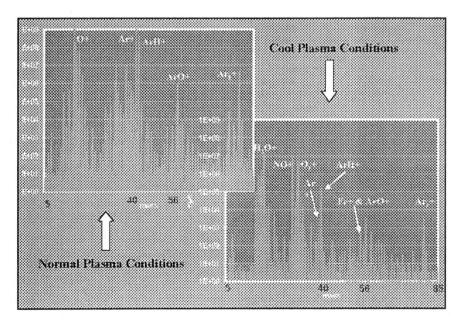

FIGURE 20.7 The intensity of the argon-based species under cool plasma conditions is significantly less than under normal plasma conditions. (From Ref. 11.)

This is typically achieved with a decrease in the RF power, an increase in the nebulizer gas flow, and sometimes a change in the sampling position of the plasma torch. Under these conditions, the formation of species such as $^{40}Ar^+$, $^{38}ArH^+$, and $^{40}Ar^{16}O^+$ are dramatically reduced, which allows for the determination of low levels of $^{40}Ca^+$, $^{39}K^+$, and $^{56}Fe^+$, respectively [12]. Under normal hot plasma conditions (typically at RF power of 1200–1600 W and a nebulizer gas flow of 0.8–1.0 L/min), these isotopes would not be available for quantitation because of the argon-based interferences. Under cool plasma conditions (typically at RF power of 600–800 W and a nebulizer gas flow of 1.2–1.6 L/min), the most sensitive isotopes can be used, offering low parts-per-trillion detection in aqueous matrices. However, not all instruments offer the same level of cool plasma performance, so if these elements are important to you, it is critical to understand what kind of detection capability is achievable. A simple way to test cool plasma performance is to look at the BEC for iron at mass 56 with respect to cobalt at mass 59. This enables the background at mass 56 to be compared to a surrogate element such as Co, which has an ionization potential similar to Fe, without actually introducing Fe into the system and contributing to the ArO background signal. When carrying out this test, it is important to use the cleanest deionized water to guarantee that there is no Fe in the blank. First measure the background in counts per second at mass 56 by aspirating deionized water. Then record the analyte intensity of a 1-ppb Co solution at mass 59. The ArO BEC can be calculated as follows:

$$BEC(ArO^+) = \frac{\text{Intensity of Deionized Water Background at Mass 56} \times 1 \text{ ppb}}{\text{Intensity of 1 ppb of Co at Mass 59} - \text{Background at Mass 56}}$$

The ArO^+ BEC at mass 56 will be a good indication of the detection limit for $^{56}Fe^+$ under cool plasma conditions. The BEC value will typically be about an order of magnitude greater than the detection limit.

Although most instruments offer cool plasma capability, there are subtle differences in the way it is implemented. It is therefore very important to evaluate the ease of setup and how easy it is to switch from cool to normal plasma conditions and back in an automated multielement run. In addition, remember that there will be an equilibrium time in switching from normal to cool plasma conditions. Make sure you know what this is because an equivalent read delay will have to be built into the method, which could be an issue if speed of analysis is important to you. If in doubt, set up a test to determine the equilibrium time by carrying out a short stability run while switching back and forth between normal and cool plasma conditions.

It is also critical to be aware that the electrical characteristics of a cool plasma are different from a normal one. This means that unless there is a good grounding mechanism between the plasma and the RF coil, a secondary discharge can easily occur between the plasma and the sampler cone. The result is an increased spread in the kinetic energy of the ions entering the mass spectrometer, making them more difficult to control and steer through the ion optics into the mass analyzer. So understand how this grounding mechanism is implemented and whether any hardware changes need to be made when going from cool to normal plasma conditions and vice versa (testing for a secondary discharge will be discussed later).

It should be noted that one of the disadvantages of the cool plasma approach is that it contains much less energy than a normal high-temperature plasma. As a result, elemental sensitivity for the majority of elements is severely affected by the matrix, which basically precludes its use for the analysis of samples with a real matrix, unless the necessary steps are taken. This is exemplified in Figure 20.8, which shows cool plasma sensitivity for a selected group of elements in varying concentrations of nitric acid, and Figure 20.9, which shows the same group of elements under hot plasma conditions. It can be seen clearly in Figure 20.9 that analyte sensitivity is dramatically reduced in a cool plasma as the acid concentration is increased, whereas under hot plasma conditions, the sensitivity for most of the elements varies only slightly with increasing acid concentration [13].

In addition, because cool plasma contains much less energy than a normal plasma, chemical matrices and acids with a high boiling point are often difficult to decompose in the plasma, which has the potential to cause corrosion problems on the interface of the mass spectrometer. This is the inherent

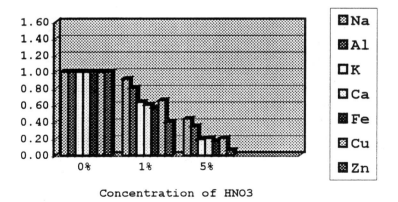

Concentration of HNO3

FIGURE 20.8 Sensitivity for a selected group of elements in varying concentrations of nitric acid, using cool plasma conditions (RF power, 800 W; nebulizer gas, 1.5 L/min). (From Ref. 13.)

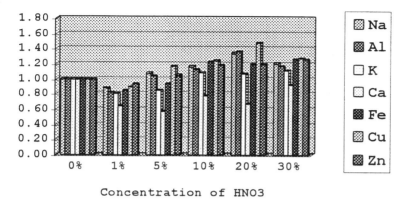

FIGURE 20.9 Sensitivity for a selected group of elements in varying concentrations of nitric acid, using hot plasma conditions (RF power, 1600 W; nebulizer gas, 1.0 L/min). (From Ref. 13.)

weakness of the cool plasma approach—instrument performance is very dependent on the sample being analyzed. As a result, unless simple aqueous-type samples are being analyzed, cool plasma operation often requires the use of standard additions or matrix matching to achieve satisfactory results. Additionally, to obtain the best performance for a full suite of elements, a multielement analysis often necessitates the use of two sets of operating conditions—one run for the cool plasma elements and another for normal plasma elements, which can be both time-consuming and sample-consuming.

In fact, these application limitations have led some vendors to reject the cool plasma approach in favor of collision/reaction cell technology instead. So it could be that the cool plasma capability of an instrument may not be that important if the equivalent elements are superior using the collision/reaction cell option. However, you should proceed with caution in this area because on the current evidence, not all collision/reaction cell instruments offer the same kind of performance. For some instruments, cool plasma detection limits are superior to the same group of elements determined in the collision cell mode. For that reason, an assessment of the suitability of using cool plasma conditions or collision/reaction cell technology for a particular application problem has to be made based on your sample and the vendor's recommendations.

Use of Collision/Reaction Cell Technology

Collision or reaction cells are predominantly used with conventional quadrupole mass analyzers to reduce the formation of harmful polyatomic spectral interferences. Although, in principle, they can be also be used with other types of mass analyzers, such as magnetic sector systems, up to now, there does not appear to be a real benefit to do this. The majority of quadrupole-based instruments on the market offer collision or reaction cell capability to reduce background levels for many of the argon-based and solvent-based spectral interferences, such as $^{38}ArH^+$, $^{40}Ar^+$, and $^{40}Ar^{16}O^+$, to improve detection capability for elements such as $^{39}K^+$, $^{40}Ca^+$, and $^{56}Fe^+$.

However, when comparing systems, it is important to understand how the interference reduction is carried out, what types of collision/reaction gases are used, and how the collision or reaction cell deals with the many complex side reactions that take place—reactions that can potentially generate brand new interfering species and cause significant problems at other mass regions. As described in "Use of Collision/Reaction Cell Technology," there are basically two different approaches used to reject these undesirable species. It can be done by either kinetic energy discrimination, or by mass discrimination, depending on the type of multipole and reaction gas used in the cell.

Unfortunately, higher-order multipoles such as hexapoles or octapoles have less defined mass stability boundaries than lower-order multipoles, making them less than ideal to intercept these side reactions by mass discrimination. This means that some other mechanism has to be used to reject these unwanted species. The approach that has been traditionally used is to discriminate them by kinetic energy. This is a well-accepted technique that is typically achieved by setting the collision cell potential slightly more negative than the mass filter potential. This means that the collision product ions generated in the cell, which have a lower kinetic energy as a result of the collision process, are rejected, whereas the analyte ions, which have a higher kinetic energy, are transmitted. This method works very well, but restricts their use to less reactive gases such as hydrogen and helium because of the limitations of higher-order multipoles to efficiently control the multitude of side reactions. However, some systems now offer a little more flexibility by being able to adjust the kinetic energy discrimination barrier with analytical mass. This enables them to use small amounts of highly reactive gases, which are recognized as being more efficient at reducing these kinds of polyatomic interferences.

However, the use of highly reactive gases such as ammonia and methane can lead to more side reactions and potentially more interferences unless the by-products from these side reactions are rejected. The way around this problem is to utilize a lower-order multipole, such as a quadrupole, inside the reaction/collision cell and to use it as a mass discrimination device. The advantage of using a quadrupole is that the stability boundaries are much better defined than a hexapole or an octapole, so it is relatively straightforward to operate the quadrupole inside the reaction cell as a mass or bandpass filter.

Therefore by careful optimization of the quadrupole electrical fields, unwanted reactions between the gas and the sample matrix or solvent, which could potentially lead to new interferences, are prevented. This means that every time an analyte and interfering ions enter the reaction cell, the bandpass of the quadrupole can be optimized for that specific problem and then changed on the fly for the next one [14].

When assessing the capabilities of collision and reaction cells, it is important to understand the level of interference rejection that is achievable. This will be reflected in the instrument's detection limits and BEC values. It will be dependent on the type of interference being reduced, but in the evidence published to date, it appears that systems that use highly reactive gases and discriminate by mass seem to offer more efficient reduction in background levels of species such as $^{38}ArH^+$, $^{40}Ar^+$, and $^{40}Ar^{16}O^+$ than systems that use higher-order multipoles and kinetic energy discrimination [15,16].

The other major benefit of the mass scanning approach is that the choice and flow of the reaction gas can be optimized for each application problem. This means that not only can you select the optimum reaction gas for different matrices, but you can also change it for different analyte masses. This enhances their flexibility over other approaches and offers the possibility of using them to reduce other problematical matrix/solvent-induced polyatomic interference such as $^{40}Ar^{35}Cl^+$, $^{32}S^{16}O^+$, and $^{40}Ar^{12}C^+$, in addition to the normal aqueous-based argon polyatomic overlaps. However, it should be emphasized that when you are comparing systems, it should be done with your particular analytical problem in mind. In other words, evaluate the BEC and detection limit performance for the suite of elements and matrices you are interested in. This will give you a very good indication of the background reduction capability of the collision/reaction cell technology you are evaluating.

Reduction of Matrix-Induced Interferences

As discussed in Chap. 14 on "Interferences," there are three major sources of matrix-induced problems in ICP-MS. The first and simplest source to overcome is often called a sample transport or viscosity effect, and is a physical suppression of the analyte signal brought on by the matrix components. It is caused by the sample's impact on droplet formation in the nebulizer or droplet size selection in the spray chamber. In some matrices, it can also be caused by the variation in sample flow through the peristaltic pump. The second type of signal suppression is caused by the impact of the sample matrix on the ionization temperature of the plasma discharge. This is typically exemplified when different levels of matrix components or acids are aspirated into a cool plasma. The ionization conditions in the low-temperature plasma are so fragile that higher concentrations of matrix components result in severe suppression of the analyte signal. The third major cause of matrix suppression is the result of poor transmission of ions through the ion optics due to matrix-induced space charge effects [17]. This has the effect of defocusing the ion beam, which leads to poor sensitivity and detection limits, especially when trace levels of low mass elements are determined in the presence of large concentrations of high mass matrices. Unless an electrostatic compensation is made in the ion optical region, the high mass element will dominate the ion beam, resulting in severe matrix suppression on the lighter ones. All these types of matrix interferences are compensated to varying degrees by the use of internal standardization, where the intensity of a spiked element that is not present in the sample is monitored in samples, standards, and blanks.

The single biggest difference in the approach of commercial ICP-MS systems to steer the maximum number of analyte ions into the mass analyzer and minimize matrix-induced suppression is in the design of the ion lens system. Although they all basically do the same job of transporting the maximum number of analyte ions through the system, there are many different ways of implementing this fundamental process, including the use of extraction lens, multicomponent lens systems, dynamically scanned single ion lens, right-angled reflectors, or multipole ion guide systems. First of all, it is important to know how many lens voltages have to be optimized. If a system has many lens components, it is probably going to be more complex to carry out optimization on a routine basis. In addition, the cleaning and maintenance of a multicomponent lens system might be a little more time-consuming. All of these are possible concerns, especially in a routine environment, where maybe the skill level of the operator is not so high.

However, the design of the ion focusing system, or the number of lens components used is not as important as its ability to handle real-world matrices [18]. Most lens systems can perform in a simple aqueous sample because there are relatively few matrix ions to suppress the analyte ions. The test of the ion optics comes when samples with a real matrix are encountered. When a large number of matrix ions are present in the system, they can physically "knock" the analyte ions out of the ion beam. This shows itself as a suppression of the analyte ions, which means that less analyte ions are transmitted to the detector in the presence of a matrix. For this reason, it is important to measure the degree of matrix suppression of the instrument being evaluated, across the full mass range. The best way to do this is to choose three or four of your typical analyte elements, spread across the mass range (e.g., $^7Li^+$, $^{63}Cu^+$, $^{103}Rh^+$, and $^{138}Ba^+$). Run a calibration of a 20-ppb multielement standard in 1% HNO_3. Then make up a synthetic sample of 20 ppb of the same elements in one of your typical matrices. Read off this sample against the original calibration.

The percentage matrix suppression at each mass can then be calculated as follows:

$$\frac{20 \text{ ppb} - \text{Apparent Concentration of 20 ppb of Analytes in Your Matrix}}{20 \text{ ppb}} \times 100$$

There is a strong possibility that your own samples will not really test the matrix suppression performance of the instrument, particularly if they are simple aqueous-type samples. If this is the case and you really would like to understand the matrix capabilities of your instrument, then make up a synthetic sample of your analytes in 500 ppm of a high mass element such as thallium, lead, or uranium. For this test to be meaningful, you should tell the manufacturers to set up the ion optical voltages that are best suited for multielement analysis across the full mass range. If the ion optics are designed correctly for minimum matrix interferences, it should not matter if it incorporates an extraction lens, uses a photon stop, has an off-axis mass analyzer, or utilizes a single, multicomponent, or right-angled ion lens system.

It is also important to understand that an additional roll of the ion optical system is to stop particulates and neutral species making it through to the detector and increasing the noise of the background signal. This will certainly impact the instrument's detection capability in the presence of complex matrices. For this reason, it is definitely worth carrying out a detection limit test in a difficult matrix such as lead or uranium, which tests the ability of the ion optics to transport the maximum number of analyte ions while rejecting the maximum number of matrix ions, neutral species, and particulates.

Another aspect of an instrument's matrix capability is its ability to aspirate many different types of samples, using both conventional nebulization and sampling accessories that generate a dry aerosol, such as laser ablation or electrothermal vaporization (ETV) sampling. When changing sample types like this on a regular basis, parameters such as RF power, nebulizer gas flow, and sampling depth usually have to be changed. When this is done, there is an increased chance of altering the electrical characteristics of the plasma and producing a secondary discharge at the interface. All instruments should be able to handle this to some extent, but depending on how they compensate for the increase in plasma potential, parameters might need to be reoptimized because of the change in the spread of kinetic energy of the ions entering the mass spectrometer [19]. This may not be such a serious problem, but once again, it is important that you are aware of this, especially if the instrument is running many different sample matrices on a routine basis.

Some of the repercussions of a secondary discharge, including increased doubly charged species, erosion of material from skimmer cone, shorter lifetime of sampler cone [20], significantly different full mass range response curve with laser ablation [21], and occurrence of two signal maxima when optimizing nebulizer gas flow have been well reported in the literature [22]. On the other hand, systems that do not show signs of these phenomena have reported an absence of these deleterious effects [23].

A simple way of testing for the possibility of a secondary discharge is to aspirate one of your typical matrices containing approximately 1 ppb of a small group of elements across the mass range (such as $^7Li^+$, $^{115}In^+$, and $^{208}Pb^+$) and continuously monitor the signals while changing the nebulizer gas flow. In the absence of a secondary discharge, all three elements, with widely different masses and ion energies, should track each other and have similar optimum nebulizer gas flows. This can be seen in Figure 20.10, which shows the signals for $^7Li^+$, $^{115}In^+$, and $^{208}Pb^+$ changing as the nebulizer gas flow is changed.

If the signals do not track each other, or there is an erratic behavior in the signals, it could indicate that the normal kinetic energy of the ions has been altered by the change in the nebulizer gas flow. There are many reasons for this kind of behavior, but it could point to a possible secondary discharge at

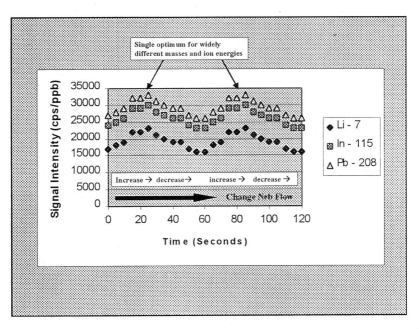

FIGURE 20.10 As the nebulizer gas flow is changed, the signals for 1 ppb of $^7Li^+$, $^{115}In,^+$ and $^{208}Pb^+$ should all track each other and have similar optimum values, if the interface is grounded correctly.

the interface, or that the RF coil grounding mechanism is not working correctly [24]. Figure 20.10 is just a graphical representation of what the relative signals might look like and might not exactly reflect all instruments. However, it should be emphasized that the difference in intensities of the elements across the mass range will also indicate the flatness of the mass response curve. In other words, the closer the intensities are to each other, the flatter the mass response curve will be. This translates into less mass discrimination, and therefore is easier to compensate for suppression effects using internal standardization.

Sample Throughput

In laboratories where high-sample throughput is a requirement, the overall cost of analysis is a significant driving force as to what type of instrument is purchased. However, in a high-workload laboratory, there sometimes has to be a compromise between the number of samples analyzed and the detection limit performance required. For example, if the laboratory wants to analyze as many samples as possible, relatively short integration times have to be used for the suite of elements being determined. On the other hand, if detection limit performance is the driving force, longer integration times need to be used, which will significantly impact the total number of samples that can be analyzed. This was described in detail in Chap. 12 on "Measurement Protocol," but it is worth revisiting to understand the full implications of achieving high-sample throughput.

It is generally accepted that for a fixed integration time, peak hopping will always give the best detection limits. As discussed earlier, measurement time is a combination of time spent on the peak-taking measurements (dwell time) and the time taken to settle (settling time) before the measurement is taken. The ratio of the dwell time to the overall measurement time is often called the measurement efficiency. The settling time, as we now know, does not contribute to the analytical signal, but definitely contributes to the analysis time. This means that every time the quadrupole sweeps to a mass and sits on the mass for the selected dwell time, there is also a settling time associated with it. The more points that have been selected to quantitate the mass, the longer is the total settling time and the worse is the overall measurement efficiency.

For example, let us take a scenario where 20 elements need to be determined in duplicate. For argument's sake, let us use an integration time of 1 sec per mass, comprising of 20 sweeps of 50 msec per sweep. The total integration time that contributes to the analytical signal and the detection limit is therefore 20 sec per replicate. However, every time the analyzer is swept to a mass, the associated scanning and settling times must be added to the dwell time.

The more points that are taken to quantify the peak, the more settling time must be added. For this scenario, let us assume that three points per peak are being used to quantify the peaks. Let us also assume for this case that the quadrupole and the detector have a settling time of 5 msec. This means that a 15-msec settling time will be associated with every sweep of each individual mass. So for 20 sweeps of 20 masses, this is equivalent to 6 sec of nonanalytical time in every replicate, which translates into 12 sec (plus 40 sec of actual measurement time) for every duplicate analysis. This is equivalent to a $40/(12+40)\times100\%$ or a 77% measurement efficiency cycle. It does not take long to realize that the fewer points taken per peak and the shorter the settling time is, the better is the measurement cycle. Just by reducing the number of points to one per peak and cutting the detector settling time by two, the nonanalytical time is reduced to 4 sec, which is a $40/(2+40)\times100\%$ or a 95% measurement efficiency per duplicate analysis. It is therefore very clear that the measurement protocol has a big impact on the speed of analysis and the number of the samples that can be analyzed in a given time. For that reason, if sample throughput is important, you should understand how peak quantitation is carried out on each instrument.

The other aspect of sample throughput is the time it takes for the sample to be aspirated through the sample introduction system into the mass spectrometer, reach a steady state signal, and then be washed out when the analysis is complete. The wash-in and wash-out characteristics of the instrument will most definitely impact its sample throughput capabilities. For that reason, it is important you know what these times are for the system you are evaluating. You should also be aware that if the instrument uses a computer-controlled peristaltic pump to deliver the sample to the nebulizer and spray chamber, it can be speeded up to reduce the wash-in and wash-out times. So this should also be taken into account when evaluating the memory characteristics of the sample introduction system.

Therefore, if speed of analysis is important to your evaluation criteria, it is worth carrying out a sample throughput test. Choose a suite of elements that represents your analytical challenge. Assuming you are also interested in achieving good detection capability, let the manufacturer set the measurement protocol (integration time, dwell time, settling time, number of sweeps, points per peak, sample introduction, wash-in/wash-out times, etc.) to get their best detection limits. If you are interested in measuring high and low concentrations, also make sure that the extended dynamic range feature is implemented. Then time how long it takes to achieve detection limit levels in duplicate from the time the sample probe goes into the sample to the time a result comes out on the screen or printer. If you have time, it might also be worth carrying out this test in an autosampler with a small number of your typical samples. It is important that detection limit measurement protocol is used because factors such as integration times and wash-out times can be compromised to reduce the analysis time. All the measurement time issues discussed in this section plus the memory characteristics of the sample introduction system will be fully evaluated with this kind of test.

Transient Signal Capability

The demands on an instrument to handle transient signals generated by sampling accessories such as laser ablation [25], electrothermal vaporization, [26], flow injection [27], or chromatography separation devices [28] are very different from conventional multielement analysis using solution nebulization. Because the duration of a sampling accessory signal is much shorter (typically 5–30 sec) than a continuous signal generated by a pneumatic nebulizer, it is critical to optimize the measurement time in order to achieve the best multielement signal-to-noise in the sampling time available. The magnitude of the problem can be seen in Figure 20.11, which shows the detection of a group of masses in a hypothetical transient peak. Very obviously, to get the

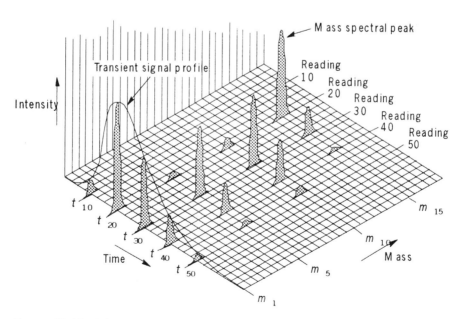

FIGURE 20.11   It is important to maximize the measurement time on a transient peak that typically lasts 2–20 sec, depending on the sampling device. (Courtesy of Perkin-Elmer Life and Analytical Sciences.)

best detection limits for this group of elements, it is important to spend all the available time quantifying the peaks of interest.

For that reason, a mass analyzer that is capable of simultaneous detection, such as a multicollector magnetic sector instrument, or at least of sampling the ions at the same time, such as the TOF design, is more desirable than a scanning analyzer, such as a single detector magnetical system, or a quadrupole-based instrument.

However, a scanning system can achieve good performance on a transient peak if the measurement time is maximized to get the best multielement signal-to-noise. For this reason, instruments that utilize short settling times are more advantageous because they achieve a higher measurement efficiency cycle. In addition, if the extended dynamic range is used to determine higher concentrations, the scanning and settling times of the detector will also have an impact on the quality of the signal. For that reason, detectors that require two scans to characterize an unknown sample will use up valuable time in the quantitation process. For example, if the transient peak is generated by an ETV sampling accessory, which only lasts 2 sec, a survey or prescan of 1 sec uses up to 50% of the available measurement time. This, of course, is a disadvantage when doing multielement analysis on a transient signal, especially if you have limited knowledge of the analyte concentration levels in your samples.

USABILITY ASPECTS

In most applications, analytical performance is a very important consideration when deciding what instrument to purchase. However, the vast majority of instruments being used today are being operated by technician level chemists. They usually have had some experience in the use of trace element techniques such as atomic absorption (AA) or ICP-OES, but in no way could be considered experts in ICP-MS. For that reason, usability aspects might be competing with analytical performance as the most important selection criterion, particularly if the application does not demand the ultimate in detection capability. Even though usability is in the eye of the user, there are some general issues that need to be addressed. They include, but are not limited to:

Software ease of use
Routine maintenance
Compatibility with sampling accessories
Installation requirements
Technical support
Training.

Software Ease of Use

First of all, you need to determine the skill level of the operator who is going to run the instrument. If it is a Ph.D.-type chemist, then maybe it is not critical that the instrument is easy to use. But if the instrument is going to be used in a high-workload environment and possibly operated round-the-clock, there is a strong possibility that the operators will not be highly skilled. For this reason, you should be looking at how easy the software is to use, and how familiar is it to other trace element techniques that are used in your laboratory. This will definitely have an impact on the time it takes to get a person fully trained on the instrument. Another issue to consider is whether the person who runs the instrument on a routine basis is the same person who will be developing the methods. Correct method development is critical because it impacts the quality of your data and, for that reason, is usually more complicated and requires more expertise than just running routine methods. I am not going to get into software features or operating systems because it is a complicated criterion to evaluate and decisions tend to be made more on a personal preference or comfort level than on the actual functionality of ICP-MS software features. However, there are differences in the way software feels. For example, if you have come from an MS background, you are probably comfortable with fairly complex research-type software. Alternatively, if you have come from a trace element background and have used AA or ICP-OES, you are probably used to more routine software that is relatively easy to use. You will find that different vendors have come to ICP-MS from a variety of different analytical chemistry backgrounds, which is often reflected in the way they design their software. Depending on the way the instrument will be used, an appropriate amount of time should be spent looking at software features that are specific to your application needs. For example, if you are a high-throughput environmental laboratory, you should be looking very closely at all the features of the automated "quality control" software, or if you do not want to spend the time to export your data to an external spreadsheet in order to create reports, you might be more interested in software with comprehensive reporting capabilities. Alternatively, if your laboratory needs to characterize lots of unknown samples, you should carefully examine the "Semiquant" software and fully understand the kind of accuracy you can expect to achieve.

Routine Maintenance

ICP mass spectrometers are complex pieces of equipment that, if not maintained correctly, will fail when you least expect them to. For that reason, a major aspect of instrument usability is how often routine maintenance has to carried out, especially if complex samples are being analyzed. You must not lose sight of the fact that your samples are being aspirated into the sample introduction system and the resulting ions generated in the plasma are steered into the mass analyzer, via the interface and ion optics. In other words, the sample, in one form or another, is in contact with many components inside the instrument. So it is essential to find out what components need to be changed and at what frequency, in order to keep the instrument in good working order. Routine maintenance has been covered in great depth in Chap. 16, but you should be asking the vendor what needs to be changed or inspected on a regular basis and what type of maintenance should be done on daily, weekly, monthly, or yearly intervals. Some typical questions might include:

> If a peristaltic pump is being used to deliver the sample, how often should the tubing be changed?
>
> How often should the spray chamber drain system be checked?
>
> Can components be changed if a nebulizer gets damaged or blocked?
>
> Can the torch sample injector be changed without discarding the torch?
>
> How is a neutral plasma maintained and if external shield or sleeves are used for grounding purposes, how often do they last?
>
> Is the RF generator solid state or does it us a power amplifier (PA) tube? (This is important because PA tubes are expensive, consumable items that typically need replacement every 1–2 years.)
>
> How often do you need to clean the interface cones and what is involved in cleaning them and keeping the cone orifices free of deposits?
>
> How long do the cones last?
>
> Do you have a platinum cone trade-in service and what is its trade-in value?
>
> What type of pump is used on the interface and if it is a rotary-type pump, how often should the oil be changed?
>
> What mechanism is used to keep the ion optics free of sample particulates or deposits?
>
> How often should the ion optics be cleaned?
>
> What is the cleaning procedure for the ion optics?
>
> Do the turbomolecular pumps require any maintenance?
>
> How long do the turbomolecular pumps last?
>
> Does the mass analyzer require any cleaning or maintenance?
>
> How long does the detector last and how easy is it to change?
>
> What spare parts do you recommend to keep on hand? (This can often indicate the components that are prone to fail most frequently.)

This is not an exhaustive list, but it should give you a good idea as to what is involved to keep an instrument in good working order. I also encourage you to talk to real-world users of the equipment to make sure you get their perspective of these maintenance issues.

Compatibility with Sampling Accessories

Sampling accessories are becoming more necessary as ICP-MS is being utilized to analyze more complex sample types. For this reason, it is important to know if the sampling accessory is made by the ICP-MS instrument company or by a third-party vendor. Obviously, if it has been made by the same company, compatibility should not be an issue. However, if it is made by a third party, you will find that some sampling accessories work much better with some instruments than they do with others. It might be that the physical connection of coupling the accessory to the ICP-MS torch has been better thought out, or that the software "talks" to one system better than another. You should refer to Chapter 17 on sampling accessories for more details on the suitability for your application, but if they are required, compatibility should be one of your evaluation objectives.

Installation of Instrument

Installation of an instrument and where it is going to be located do not seem obvious evaluation objectives at first, but could be important, particularly if space is limited. For example, is the instrument freestanding or bench-mounted because maybe you have a bench available, but no floor space or vice versa? It could be that the instrument requires a temperature-controlled room to ensure good stability and mass calibration. If this is the case, have you budgeted for this kind of expense? If the instrument is being used for ultra-trace detection levels, does it need to go into a class 1, 10, or 100? If it does, what is the size of the room and do the roughing pumps need to be placed in another room? In other words, it is important to fully understand the installation requirements for each instrument being evaluated and where it will be located. Refer to Chap. 15 on "Contamination Issues" for more information on instrument installation.

Technical Support

Technical and application support is a very important consideration, especially if you have had no previous experience with ICP-MS. You want to know that you are not going to be left on your own after you have made the purchase. For this reason, it is important not only to know the level of expertise of the specialist who is supporting you, but also whether they are local to you or located in the manufacturer's corporate headquarters. In other words, can you guarantee getting technical help whenever you need it? Another important aspect related to application support is the availability of application literature. Is there a wide selection of materials available for you to read, either in the form of web-based application reports or references in the open literature, to help you develop your methods? In addition, find out if there are active user or Internet-based discussion groups because they will be invaluable sources of technical and application help.

Training

Find out what kind of training course comes with the purchase of the instrument and how often it is run. Most instruments come with a 2- to -3-day training course for one person, but most vendors should be flexible on the number of people who can attend. Some manufacturers also offer application training where they teach you how to optimize methods for major application areas such as environmental, clinical, and semiconductor analyses. Talk to other users about the quality of the training they received when they purchased their instruments and also ask them what they thought of the operator manuals. You will often find that this is a good indication of how important a manufacturer views customer training.

RELIABILITY ISSUES

To a certain degree, instrument reliability is impacted by routine maintenance issues and the types of samples being analyzed, but it is generally considered more of a reflection of the design of an instrument. Most manufacturers will guarantee a minimum percentage uptime for their instrument, but this number (which is typically ~95%) is almost meaningless unless you really understand how it is calculated. Even when you know how it is calculated, it is still difficult to make the comparison, but at least you should understand the implications if the vendor fails to deliver. Good instrument reliability is taken for granted nowadays, but it has not always been the case. When ICP-MS was first commercialized, the early instruments were a little unpredictable, to say the least, and were quite prone to frequent breakdowns. But as the technique became more mature, the quality of instrument components got better and, as a result, the reliability improved. However, you should be aware that there are components of the instrument that are more problematical than others. This is particularly true when the design of an instrument is new, or a model has had a major redesign. You will therefore find that in the life cycle of a newly designed instrument, the early years will be more susceptible to reliability problems than when the instrument is of a more mature design.

When we talk about instrument reliability, it is important to understand whether it is related to the samples being analyzed, the lack of expertise of the person operating the instrument, an unreliable component, or maybe just an inherent weakness in the design of the instrument. For example, how does the instrument handle highly corrosive chemicals, such as concentrated mineral acids? Some sample introduction systems and interfaces will be more rugged than others and require less maintenance in this area. On the other hand, if the operator is not aware of the dissolved solids limitation of the instrument, they might attempt to aspirate a sample that will slowly block the interface cones, causing signal drift and, in the long term, possible instrument failure. Or it could be something as unfortunate as a major component such as the RF generator power amplifier tube, dynode detector, or turbomolecular pump (which all have a finite lifetime) failing in the first year of use.

Service Support

Instrument reliability is very difficult to assess at the evaluation stage, so for this reason, you have to look very carefully at the kind of service support offered by the manufacturer. For example, how close is a qualified support engineer to you, or what is the maximum amount of time you will have to wait to get a support engineer at your laboratory, or at least to call you back to discuss the problem? Ask the vendor if they have the capability for remote diagnostics, where a service engineer can remotely run the instrument or check the status of a component by "talking to" your system computer via a modem. Even if this approach does not fix the problem, at least the service engineer can arrive at your laboratory with a very good indication of what it could be.

You should know up-front what it is going to cost for a service visit, irrespective of what component has failed. Most companies charge an hourly rate for a service engineer (which typically includes travel time as well), but if an overnight stay is required, fully understand what you are paying for (accommodation, meals, gas, etc). Some companies might even charge for mileage between the service engineer's base and your laboratory. Moreover, if you are a commercial laboratory and cannot afford the instrument to be down for any length of time, find out what it is going to cost for 24/7 service coverage.

You can take a chance and just pay for each service visit, or you might want to budget for an annual preventative maintenance contract, where the service engineer checks out all the important instrumental components and systems on a frequent basis to make sure they are all working correctly. This might not be as critical if you work in an academic environment, where the instrument might be down for extended periods, but in my opinion, it is absolutely critical if you are a commercial laboratory that is using the instrument to generate revenue. Find out what is included in the contract because some will also cover the cost of consumables and/or replacement parts, whereas others just cover the service visits. These annual preventive maintenance contracts are typically about 10–15% of the cost of the instrument, but are well worth it if you do not have the expertise in-house, or you just feel more comfortable with having an "insurance policy" to cover instrument breakdowns.

Once again, talking to existing users will give you a very good perspective of the quality of the instrument and/or the service support offered by the manufacturer. There is no absolute guarantee that the instrument of choice is going to perform to your satisfaction 100% of the time, but if you are a high-throughput, routine laboratory, make sure it will be down for the minimum amount of time. In other words, fully understand what it is going to cost you to maximize the uptime of all the instruments being evaluated.

FINANCIAL CONSIDERATIONS

The financial side of choosing an ICP mass spectrometer can often dominate the selection process; that is, if you have not budgeted quite enough money to buy a top-of-the-line instrument, or perhaps you had originally planned to buy another lower-cost trace element technique, or you could be using funds left over at the end of your financial year. All these scenarios could dictate how much money you have available and what kind of instrument you can purchase. In my experience, you should proceed with caution in this kind of situation because if only one manufacturer is willing to do a deal with you, the evaluation process will be a waste of time. For this reason, you should budget at least 12 months before you are going to make a purchase and add another 10–15% for inflation and any unforeseen price increases. In other words, if you want to get the right instrument for your application, never let price be the overriding factor in your decision. Always be wary of the vendor who will undercut everyone else to get your business. There could be a very good reason why they are doing this, such as that the instrument is being discontinued for a new model, or it could be having some reliability problems that are affecting its sales.

This is not to say that price is unimportant, but what might appear to be the most expensive instrument to purchase might be the least expensive to run. For that reason, you must never forget the cost of ownership in the overall financial analysis of your purchase. So by all means, compare the price of the instrument, computer, and any accessories you buy, but also factor in the cost of consumables, gases, and electricity based on your usage. Maybe instrument consumables from vendor A are much less expensive than vendor B, or maybe you can analyze far more samples with instrument A because it does not drift as much as instrument B and therefore does not need recalibrating as often. It also follows that if you can get through your daily allocation of samples much faster with one instrument than another, then your argon consumption will be less.

Another aspect that should be taken into consideration is the salary of the operator. Even though you might think that this is a constant, irrespective of the instrument, you must assess the expertise required to run it. For example, if you are thinking of purchasing more complex technology such as a magnetical sector instrument for a research-type application, the operator needs to be of a much higher skill level than, say, someone who is being asked to run a routine application with a quadrupole-based instrument. As a result, the salary of that person will probably be higher.

Finally, if one instrument has to be installed in a temperature-controlled, air-conditioned environment for stability purpose, the cost of preparing or building this kind of specialized room must be taken into consideration when doing your financial analysis. In other words, when comparing systems, never automatically reject the instrument that is the most expensive. You will find that over the 10 years that you own the instrument, the cost of doing analysis and the overall cost of ownership are more important evaluation criteria.

SUMMARY OF THE EVALUATION PROCESS

As mentioned earlier in this chapter, it was not my intention to compare instrument designs and features, but to give you some general guidelines as to what are the most important evaluation criteria, based on my experience as a product and application specialist for a manufacturer of ICP-MS equipment. Besides being a framework for your evaluation process, these guidelines should also be used in conjunction with the other chapters in this book and with the reference information available in the public domain.

But if you want to find the best instrument for your application needs, be prepared to spend a few months evaluating the marketplace. Do not forget to prioritize your objectives and give each of them a weighting factor, based on their degree of importance for the types of samples you analyze. Be careful to take the evaluation in a direction you want to take it and not where the vendor wants to. In other words, it is important to compare apples with apples. However, be prepared that there might not be a clear-cut winner at the end of the evaluation. If this is the case, then decide what aspects of the evaluation are most important and ask the manufacturer to put them in writing. Some vendors might be hesitant to do this, especially if it is an instrument performance issue.

Talk to as many users in your field as you possible can—not only ones given to you by the vendor, but ones chosen by yourself also. This will give you a very good indication as to the real-world capabilities of the instrument, which can often be overlooked at a demonstration. You might find, from talking to "typical" users, that it becomes obvious which instrument to purchase. If that is the case and your organization allows it, ask the vendor what kind of deal they can give you if you do not have samples to run and you do not want a demonstration. I guarantee you will be in a much better position to negotiate a lower price.

Never forget that it is a very competitive marketplace and your business is extremely important to each of the ICP-MS manufacturers. Hopefully, this book has not only helped you understand the fundamentals of the technique a little better, but has also given you some thoughts and ideas as how to find the best instrument for your needs. Good luck.

FURTHER READING

1. Newman A. Elements of ICP-MS: product review. Anal Chem, January 1996, 46A–51A.
2. Royal Society of Chemistry. Report by the Analytical Methods Committee: evaluation of analytical instrumentation: Part X. Inductively coupled plasma mass spectrometers. Analyst 1997; 122:393–408.
3. Montasser A, ed. Inductively Coupled Plasma Mass Spectrometry: An Introduction to ICP Spectrometries for Elemental Analysis—Analytical Figures of Merit for ICP-MS. Chap. 1.4. Berlin: Wiley-VCH, 1998:16–28.
4. Denoyer ER. At Spectr 1992; 13(3):93–98.
5. Thomsen MA. At Spectr 2000; 13(3):93–98.
6. Halicz L, Erel Y, Veron A. At Spectr 1996; 17(5):186–189.
7. Thomas R. Spectroscopy 2002; 17(7):44–48.
8. Denoyer ER, Lu QH. At Spectr 1993; 14(6):162–169.
9. Hutton R, Walsh A, Milton D, Cantle J. CHEMSA 1991; 17:213–215.
10. Dawson PH, ed. Quadrupole Mass Spectrometry and Its Applications. Amsterdam: Elsevier, 1976. reissued by AIP Press, Woodbury, NY, 1995.
11. Jiang SJ, Houk RS, Stevens MA. Anal Chem 1988; 60:217.
12. Sakata K, Kawabata K. Spectrochim Acta 1994; 49B:1027.
13. Collard JM, Kawabata K, Kishi Y, Thomas R. Micro, January 2002; 2(1):39–46.
14. Tanner SD, Baranov VI. At Spectr 1999; 20(2):45–52.
15. Feldman I, Jakubowski N, Thomas C, Stuewer D. Fresnius J Anal Chem 1999; 365:422–428.
16. Voellkopf U, Klemm K, Pfluger M. At Spectr 1999; 20(2):53–59.
17. Tanner SD, Douglas DJ, French JB. Appl Spectrosc 1994; 48:1373.
18. Denoyer ER, Jacques D, Debrah E, Tanner SD. At Spectr 1995; 16(1):1.
19. Hutton RC, Eaton AN. J Anal At Spectrom 1987; 5:595.

20. Gray AL, Date A. Analyst 1981; 106:1255.
21. Wyse EJ, Koppenal DW, Smith MR, Fisher, DR. 18th FACSS Meeting, Anaheim, CA, October, 1991, Paper No. 409.
22. Diegor WG, Longerich HP. At Spectr 2000; 21(3):111.
23. Douglas DJ, French JB. Spectrochim Acta 1986; 41B(3):197.
24. Denoyer ER. At Spectr 1991; 12:215–224.
25. Denoyer ER, Fredeen KJ, Hager JW. Anal Chem 1991; 63(8):445–457.
26. Beres SA, Denoyer ER, Thomas R, Bruckner P. Spectroscopy 1994; 9(1):20–26.
27. Stroh A, Voellkopf U, Denoyer E. J Anal At Spectrom 1992; 7:1201.
28. Ebdon L, Fisher A, Handley H, Jones P. J Anal At Spectrom 1993; 8:979–981.

21

Useful Contact Information

The final chapter of the book is dedicated to providing you with useful contact information related to ICP-MS. It is not exhaustive by any means, but includes information about alternative sources of sample introduction consumables and other instrument components; suppliers of laboratory chemicals, calibration standards, certified reference materials, high-purity gases, deionized water systems, and clean room equipment. I have also included information about the major scientific conferences, professional societies, publishing houses, Internet discussion groups, and the most popular ICP-MS-related journals. I hope you find it useful.

Certified reference materials/calibration standards	
National Research Council of Canada 1500 Montreal Road Ottawa, Ontario, K1A 0R9, Canada Phone: 800-668-1222 Fax: 613-952-8239 www.nrc.ca	NIST 100 Bureau Drive, Stop 200 Gaithersburg, MD 20899 Phone: 301-975-6776 Fax: 301-975-2183 www.nist.gov
Conostan Inc. 1000 S. Pine, P.O. Box 1267 Ponca City, OK 74602 Phone: 580-767-3078 Fax: 580-767-5843 www.conostan.com	High Purity Standards P.O. Box 41727 Charleston, SC 29423 Phone: 843-767-7900 Fax: 843-767-7906 www.hps.net

Inorganic Ventures Inc.
7000 High Grove Boulevard
Burr Ridge, IL 60521
195 Lehigh Avenue, Suite 4
Lakewood, NJ 08701
Phone: 800-669-6799
Fax: 732-901-1903
www.ivstandards.com

SPEX Certiprep
6141 Easton Road, P.O. Box 310
Plumsteadville, PA 18949
203 Norcross Avenue
Metuchen, NJ 08840
Phone: 800-522-7739
Fax: 732-603-9647
www.spexcsp.com

Chemicals/Standards

Aldrich Chemicals
940 W. St. Paul Avenue
Milwaukee, WI 53233
Phone: 414-273-3850
Fax: 414-273-4979
www.sigma-aldrich.com

Eichrom Technologies Inc.
8205 S. Cass Avenue
Darien, IL 60561
Phone: 800-422-6693
Fax: 630-963-1928
www.eichrom.com

Fisher Scientific Inc.
2000 Park Lane
Pittsburgh, PA 15275
Phone: 412-490-8472
Fax: 412-809-1310
www.fishersci.com

J. T. Baker
222 Red School Lane
Phillipsburg, NJ 08865
Phone: 908-859-9315
Fax: 908-859-9385
www.jtbaker.com

Clean room equipment

Cleanroom Consulting LLC
5396 Springview Drive
Fayetteville, NY 13066
Phone: 315-637-4030
Fax: 315-637-0928
www.cleanroomconsulting.com

Clestra Cleanroom Inc.
7000 Performance Drive N
Syracuse, NY 13212
Phone: 315-452-5200
Fax: 315-452-5252
E-mail: clestraus@aol.com

Microzone Corp.
25F Northside Road, PO Box 11336
Ottawa, Ontario, Canada K2H 7V1
Phone: 613-829-1433
Fax: 613-829-6331
www.microzone.com

Consumables (sample introduction/interface)

Burgener Research Inc.
1680–2 Lakeshore Road W.
Mississauga, Ontario,
Canada L5J 1J5
Phone: 905-823-3535
Fax: 905-823-2717
www.burgenerresearch.com

Sherba Analytical Inc.
P. O. Box 880
New Port Richey,
FL 34652
Phone: 800-228-5085
Fax: 727-844-3613
www.sherba.com

CPI International
5580 Skylane Boulevard
Santa Rosa, CA 95403
Phone: 800-878-7654
Fax: 707-545-7901
www.cpiinternational.com

Elemental Scientific Inc (ESI)
2440 Cumming Street
Omaha, NE 68131
Phone: 402-991-7800
Fax: 402-997-7799
www.elementalscientific.com

Glass Expansion Pty.
15 Batman Street
West Melbourne,
Victoria 3003, Australia
Phone: 61-3-9320-111
Fax: 61-3-9320-1112
www.geicp.com

Meinhard Glass Products
700 Corporate Circle, Suite A
Golden, CO 80401
Phone: 303-277-9776
Fax: 303-216-2649
www.meinhard.com

Precision Glassblowing
14775 E. Hindsdale Avenue
Engelwood, CO 80112
Phone: 303-693-7329
Fax: 303-699-6815
www.precisionglassblowing.com

SCP Science
21800 Clark Graham
Baie D'urfe, Canada H9X 4B6
Phone: 800-361-6820
Fax: 514-457-4499
www.scpscience.com

Spectron Inc.
2080 Sunset Drive
Ventura, CA 93001
Phone: 805-652-1992
Fax: 805-652-1994
www.spectronus.com

Superior Glassblowing Co.
7190 Vreeland Road
Ypsilanti, MI 43198
Phone: 734-482-8744
Fax: 734-482-0672
Email:sgb5581@aol.com

Consumables (detectors)

SGE Inc.
2007 Kramer Lane
Austin, TX 78758
Phone: 800-945-6254
Fax: 512-836-9159
www.ecpsci.com

Deionized water systems

Millipore Corporation
80 Ashby Road
Bedford, MA 01730
Phone: 800-645-5476
Fax: 800-645-5439
www.millipore.com

U.S. Filter
1501 E. Woodfield Road, Suite
200W, Schauburg, IL 60173
Phone: 800-466-7873
www.usfilter.com

Expositions and conferences

Eastern Analytical
P.O. Box 633
Montchanin, DE 19710
Phone: 610-485-4633
Fax: 610-485-9467
www.eas.org

FACSS
1201 Don Diego Avenue
Santa Fe, NM 87505
Phone: 505-820-1648
Fax: 505-989-1073
www.facss.org

Pittsburgh Conference (PittCon)
300 Penn Center Boulevard,
Suite 332 Pittsburgh,
PA 15235
Phone: 412-825-3220
Fax: 412-825-3224
www.pittcon.org

Winter Conference on Plasma
Spectrochemistry
c/o Dr. Ramon Barnes
85 N. Whitney Street
Amherst, MA 01002-1869
Phone: 413-256-8942
Fax: 413-256-3746
E-mail: winterconf@chem.umass.edu

Gases

Air Liquide
2700 Post Oak Boulevard
Houston, TX 77056
Phone: 800-248-1427
Fax: 281-474-8419
www.airliquide.com

Air Products and Chemicals Inc.
7201 Hamilton Boulevard
Allentown, PA 18195
Phone: 800-654-4567
Fax: 800-880-5204
www.airproducts.com

Praxair Specialty Gases
7000 High Grove Boulevard
Burr Ridge, IL 60521
Phone: 877-772-9247
Fax: 630-320-4506
www.praxair.com/specialty gases

Scott Specialty Gases
6141 Easton Road, P.O. Box 310
Plumsteadville, PA 18949
Phone: 215-766-8861
Fax: 215-766-2476
www.scottgas.com

Spectra Gases
3434 Rt. 22, West Branchburg,
NJ 08876
Phone: 800-932-0624
Fax: 908-252-0811
www.spectragases.com

Instrumentation (quadrupole technology)

Agilent Technologies
2850 Centerville Road,
Wilmington, DE 19808
Phone: 302-633-8264
Fax: 302-633-8916
www.chem.agilent.com

GV Instruments
Crewe Road
Wythenshawe, Manchester
M23 9BE, England, UK
Phone: 0161 9022100
Fax: 0161 9022198
www.gvinstruments.co.uk

PerkinElmer Life and
Analytical Sciences
710 Bridgeport Avenue
Shelton, CT 06484
Phone: 800-762-4000
Fax: 203-944-4914
www.perkinelmer.com

Thermo Elemental
27 Forge Parkway
Franklin, MA 02038
Phone: 800-229-4087
Fax: 508-528-2127
www.thermoelemental.com

Varian, Inc.
2700 Mitchell Drive
Walnut Creek,
CA 94598
Phone: 800-926-3000
Fax: 925-945-2360
www.varianinc.com

Instrumentation (magnetic sector technology)

GV Instruments
Crewe Road
Wythenshawe, Manchester
M23 9BE, England, UK
Phone: 0161 9022100
Fax: 0161 9022198
www.gvinstruments.co.uk

Thermo Finnigan
355 River Oaks Parkway
San Jose, CA 95134
Phone: 408-965-6000
Fax: 408-965-6010
www.thermofinnigan.com

Instrumentation (time-of-flight technology)

GBC Scientific
3930 Ventura Drive, Suite 350
Arlington Heights, IL 60004
Phone: 800-445-1902
Fax: 847-506-1901
www.gbcsci.com

Leco Corporation
3000 Lakeview Avenue
St. Joseph, MI 49085
Phone: 616-985-4711
Fax: 616-982-8987
www.leco.com

Internet discussion group

PLASMACHEM List Server
312 Heroy Geology Laboratory
University of Syracuse
Syracuse, NY 13244
Phone: 315-443-1261
(Michael Cheatham)
Fax: 315-443-3363
To subscribe:
E-mail: mmcheath@mailbox.syr.edu

Journals/Magazines

American Laboratory
30 Control Drive
Shelton, CT 06484
Phone: 800-777-9009
Fax: 203-926-9310
www.iscpubs.com

Analytical Chemistry
1155 16th Street, NW
Washington, DC 20036
Phone: 202-872-4570
Fax: 202-872-4574
www.pubs.acs.org/ac

ICP Information Newsletter Inc.
P. O. Box 666
Hadley, MA 01035-0666
Phone: 413-256-8942
Fax: 413-256-8942
Email: icpnews@chem.umass.edu

Spectroscopy Magazine
(Editorial Office)
859 Willamette Street
Eugene, OR 97401
Phone: 541-343-1200
Fax: 541-984-5250
www.spectroscopyonline.com

JAAS (Royal Society of Chemistry)
Thomas Graham House
Science Park, Milton Road
Cambridge, CB4 4WF, England, UK
Phone: 44-1223-420066
Fax: 44-1223-420247
www.rsc.org

Spectrochmica Acts (Part B)
c/o The Editor,
Dr. G. de Loos-Vollebregt,
Delft University of Technology,
Facility of Applied Sciences,
DCT-TOCK,
Julianalaan 136,
2628 BL Delft,
The Netherlands
Email: m.t.c.deloos@tnw.tudelft.nl

Professional societies/services

American Chemical
Society (ACS)
1155 16th Street NW
Washington, DC 20036
Phone: 800-227-5558
Fax: 202-872-4615
www.pubs.acs.org

ASTM
100 Barr Harbor Drive,
West Conshohocken, PA 19428
Phone: 610-832-9605
Fax: 610-834-3642
www.astm.org

Chemical Abstract Services (CAS)
2540 Olentangy Drive
Columbus, OH 43202
Phone: 800-753-4227
Fax: 614-447-3837
www.cas.org

American Society for Mass
Spectrometry
1201 Don Diego Avenue
Santa Fe, NM 87505
Phone: 505-989-4517
Fax: 505-989-1073
www.asms.org

Society for Applied
Spectroscopy (SAS)
201b Broadway Street
Frederick, MD 21701
Phone: 301-694-8122
Fax: 301-694-6860
www.s-a-s.org

Publishers

Marcel Dekker	Elsevier Science Publishing
270 Madison Avenue	655 Avenue of the Americas
New York, NY 10016	New York, NY 10010
Phone: 212-696-9000	Phone: 212-633-3756
Fax: 212-685-4540	Fax: 212-633-3112
www.dekker.com	www.elsevier.com
John Wiley and Sons	International Scientific
605 Third Avenue	Communications
New York, NY 10158	30 Control Drive, P.O. Box 870
Phone: 212-850-6518	Shelton, CT 06484
Fax: 212-850-6617	Phone: 800-777-9009
www.wiley.com	Fax: 203-926-9310
	www.iscpubs.com

Microwave dissolution equipment

CEM Corporation	Milestone Inc.
3100 Smith Farm Road	160 B Shelton Road
Matthews, NC 62810	Monroe, CT 06468
Phone: 800-726-3331	Phone: 203-261-6175
Fax: 704-821-5185	Fax: 203-261-6592
www.cem.com	www.milestonesci.com

Sample delivery systems (autosamplers, dilutors)

CETAC Technologies	Gilson Inc.
5600 S. 42nd Street	3000 W. Beltline Highway
Omaha, NE 68107	P. O. Box 620027
Phone: 402-733-2829	Middleton, WI 53562
Fax: 402-733-5292	Phone: 800-445-7661
www.cetac.com	Fax: 608-861-4451
	www.gilson.com

Sample delivery systems (laser ablation equipment)

CETAC Technologies	New Wave Research
5600 S. 42nd Street	47613 Warm Springs Boulevard
Omaha, NE 68107	Fremont, CA 94539
Phone: 402-733-2829	Phone: 510-249-1550
Fax: 402-733-5292	Fax: 510-249-1551
www.cetac.com	www.new-wave.com

Vacuum pumps and components

Leybold Vacuum (USA) Inc.
5700 Mellon Road
Export, PA 15632
Phone: 800-764-5369
Fax: 724-733-1217
www.leyboldvacuum.com

Index

315